Theory and Practice of Urban Sustainability Transitions

drift

GraSPP
THE UNIVERSITY OF TOKYO

This book series Theory and Practice of Urban Sustainability Transitions is intended to explore the different dynamics, challenges, and breakthroughs in accelerating sustainability transitions in urban areas across the globe. We expect to find as much different and diverse stories, visions, experiments, and creative actors as there are cities: from metropolises to country towns, from inner city districts to suburbs, from developed to developing, from monocultural to diverse, and from hierarchical to egalitarian. But we also expect to find patterns in processes and dynamics of transitions across this diversity. Transition dynamics include locked-in regimes that are challenged by changing contexts, ecological stress and societal pressure for change as well as experiments and innovations in niches driven by entrepreneurial networks, and creative communities and proactive administrators. But also included are resistance by vested interests and sunken costs, uncertainties about the future amongst urban populations, political instabilities, and the erosion of social services and systems of provision. And finally there are the forming of transformative arenas, the development of coalitions for change across different actor groups, the diffusion and adoption of new practices, and exponential growth of sustainable technologies.

For this series we seek this middle ground: between urban and transition perspectives, between conceptual and empirical, and between structural and practical. We aim to develop this series to offer scholars state-of-the-art theoretical developments applied to the context of cities. Equally important is that we offer urban planners, professionals, and practitioners interested or engaged in strategic interventions to accelerate and guide urban sustainability transition frameworks for understanding and dealing with on-going developments, methods, and instruments.

This book series will lead to new insights into how cities address the sustainability challenges they face by not returning to old patterns but by searching for new and innovative methods and instruments that are based on shared principles of a transitions approach. Based on concrete experiences, state-of-the-art research, and ongoing practices, the series provides rich insights, concrete and inspiring cases as well as practical methods, tools, theories, and recommendations. The book series, informed by transition thinking as it was developed in the last decade in Europe, aims to describe, analyse, and support the quest of cities around the globe to accelerate and stimulate such a transition to sustainability.

To sum up, the book series aims to:

– Provide theory, case studies, and contextualized tools for the governance of urban transitions worldwide
– Provide a necessary and timely reflection on current practices of how Transition Management is and can be applied in urban contexts worldwide
– Further the theorizing and conceptual tools relating to an understanding of urban sustainability transitions
– Provide best practices of cities across countries and different kinds of cities as well as across policy domains in shaping their city's path towards sustainability

More information about this series at http://www.springer.com/series/13408

Trivess Moore • Fjalar de Haan • Ralph Horne
Brendan James Gleeson
Editors

Urban Sustainability Transitions

Australian Cases- International Perspectives

Springer

Editors
Trivess Moore
School of Property, Construction
and Project Management
RMIT University
Melbourne, VIC, Australia

Fjalar de Haan
Melbourne Sustainable Society Institute
and Melbourne School of Design
The University of Melbourne
Melbourne, VIC, Australia

Ralph Horne
College of Design and Social Context
RMIT University
Melbourne, VIC, Australia

Brendan James Gleeson
Melbourne Sustainable Society Institute
and Melbourne School of Design
The University of Melbourne
Melbourne, VIC, Australia

ISSN 2199-5508 ISSN 2199-5516 (electronic)
Theory and Practice of Urban Sustainability Transitions
ISBN 978-981-10-4791-6 ISBN 978-981-10-4792-3 (eBook)
DOI 10.1007/978-981-10-4792-3

Library of Congress Control Number: 2017951733

Printed on acid-free paper

This Springer imprint is published by Springer Nature
The registered company is Springer Nature Singapore Pte Ltd.
The registered company address is: 152 Beach Road, #21-01/04 Gateway East, Singapore 189721, Singapore

Foreword

It is an often-heard and well-known critique that early yet foundational work on sustainability transitions was characterized by poor spatial sensitivity and methodological nationalism. Similar to many other fields in social science, sustainability transitions has however witnessed a spatial turn over the past years, bringing questions to the table such as why do sustainability transitions unfold unevenly across space, how does spatial context matter for transition processes and how are sustainability transitions governed at and across different spatial scales?

Through its spatial turn, work in sustainability transitions has increasingly zoomed in on the role of cities in driving pathways for transformative change across a range of sectors such as energy, transport, food and housing. Moreover, in their endeavour to better understand urban contexts, transition researchers have encountered the work and insights stemming from urban studies and human geography and vice versa. As this book bears witness to, these encounters have led to fruitful transdisciplinary collaborations as well as mutual scrutiny and critique (bearing the potential to be equally, if not even more, fruitful).

Taking us on a journey down under, this book makes an original, rich and thoughtful contribution to a better understanding of the spatialities of sustainability transitions in at least two significant ways.

First of all, it helps to tease out some of the particularities of Australian urban sustainability transitions. Without becoming idiosyncratic, there are important wider insights to be gained from studying sustainability transitions in an Australian urban context, e.g. how local government and urban governance arrangements are going against the grain of a carbon regime that is largely produced and reproduced at national levels of decision- and policy-making, how urban transitions unfold in the face of weak and fragmented formal administrative power at the urban level or how urban sustainability transitions take place under the pressure of steep population growth and increasing urbanization. While focused on Australian urban sustainability transitions, many studies explicitly adopt an international comparative perspective, allowing them to not only argue that context matters but also to suggest how.

Secondly, the book adds substantial new food for thought to conceptualizing and theorizing urban sustainability transitions. In much of the literature interrogating the spatial aspects of sustainability transitions, a geographical perspective has simply been latched onto existing frameworks such as the multi-level perspective or strategic niche management. This book sets out to move beyond that as it provides an arena for urban and transition researchers to scrutinize existing wisdoms in respective fields. For example, insights and approaches from not only critical urban theory but also urban planning meet Transition Management, providing a fertile ground for new theorizing around urban sustainability transitions. This raises attention for questions and discussions that have remained under the radar such as how do transitions to low-carbon energy systems relate to social justice and socio-economic equity or how can niche experiments around resilience building by urban communities be scaled or diffused to make systemic impact?

Besides its academic merit, this book also makes a convincing case that sustainability transitions are needed to make Australian cities future-proof and provides inspiring examples of how this can be achieved in practice through interventions in urban planning, grassroots initiatives and technological innovation and diffusion.

City of Melbourne Chair in Resilient Cities
Melbourne Sustainable Society Institute
Faculty of Architecture, Building & Planning
The University of Melbourne

Professor Lars Coenen

Series Foreword

This book series seeks to bring together the heads, hearts and hands of urban sustainability transitions across the globe. In all cities by now, citizens, civil servants, researchers, entrepreneurs and societal organizations are coming together to renew urban systems of energy, water management, housing, mobility, care and food. The Paris Agreement signifies the global political commitment to serious and deep decarbonization, but it also has placed the role of cities centre stage. While cities have always been places of experimentation and innovation, the challenge now is to radically transition our urban economies, infrastructures and socio-cultural regimes towards inclusive, just, sustainable and thriving urban communities. There is a shift therefore from places of innovation to laboratories for system innovation and transition. In this grand experiment at global urban scale, there is no one size fits all; sustainability needs to be reinvented over and over again in each neighbourhood, sector and community. At the same time, there is much to learn across experiments, locations, sectors and professions.

This specific volume brings a richness of ideas, examples and insights from Australia. It compiles years of thinking and experimentation in Australian cities by scholars, practitioners and policy-makers. It not only gives an exciting and inspiring insight into specific aspects and challenges Australian cities face but also offers a lot of concrete ways to understand, influence and accelerate urban sustainability transitions. Obviously issues that relate to water, urban sprawl and energy figure prominently. Like other modern cities, the dominant solutions are often technological, centralized and fossil based. This book describes how nonetheless alternatives are developing and gaining speed. These range from water-sensitive cities and radically new concepts for urban transformation to advancing renewable energies and dealing with incumbent actors seeking to frustrate transitions efforts.

This volume also offers diverse and thoughtful reflections upon the theory and practices of transitions. It adds to the growing literature on understanding transitions and their governance by developing new ideas around the role of space and agency in transitions, by taking a critical stance towards more pragmatic and managerial approaches to transitions and by developing new ideas around such themes. This volume can therefore also be seen as evidence for the maturation of transition research in Australia. It puts the continent firmly on the transition map and reminds the world that there is yet even more to discover in and learn from Australia.

Rotterdam, The Netherlands Derk Loorbach

Preface

The move to cities and concerns about environmental sustainability are two of the defining trends shaping the twenty-first century. This book is the third in a series about urban transitions. It contributes to debates about purposive transitions to sustainable cities through an accessible but critical exploration of key changes in urban settings. Urban environments comprise a myriad of complex and imbricated environmental, social and governance challenges (e.g. climate change, population growth, equity, resource constraints, geopolitical power shifts and rapid innovation in technology and other areas). 'Business-as-usual' market-based responses or individual top-down government interventions are increasingly recognised as inadequate responses to enable sustainable cities. As a result, urban practitioners and scholars alike are concerned with driving purposive transitions to sustainable cities.

This volume provides an alternative perspective to that of the preceding book in this series by Loorbach et al. (2016). While Loorbach et al. geographically focus on European and Japanese cases with a particular interest in the application of the transition management approach, our volume explores Australian cases and their international implications with an interest in drawing in very different perspectives from urban studies.

This book is intended to take transition scholars on a tour of the city, via an introduction to perspectives from urban studies on urban change in an era dominated by neoliberal economics and climate change. It is equally intended to address urban scholars and introduce a burgeoning and lively literature of sociotechnical transitions, where there is ample scope for both empirical and theoretical insights. Perhaps most importantly of all, it is designed to introduce a broader audience to both genres and, through this, to provide different and distinctive perspectives on urban low carbon transitions, their challenges and prospects.

While we acknowledge the burgeoning literature on geographies of transitions, this book is distinctive in that it has one foot in urban studies and the other in transition studies – both in the sense of the authors' backgrounds and the contents of their chapters. We have observed urban scholarship moving towards issues of transitions and transition studies increasingly engaging with urban considerations (as demonstrated by the previous volume). This volume presents an overview of theory, concepts, approaches and practical examples informed by sustainability transitions thinking for urban practitioners and scholars who want to improve their understanding of the prospects and pathways for sustainable urban futures. The chapters in this volume contribute to the growing literature on city-scale transformative change that seeks to address a lack of consideration of spatial and urban governance dimensions in sustainability transitions studies.

Australia is representative of many OECD countries in terms of overall economic and institutional arrangements, yet it also has some unique attributes that serve to highlight issues for urban transitions internationally. Australia is a highly urbanised country with the metropolitan capitals accounting for large shares of their state populations. The five largest, state-capital cities dominate economically and demographically; however, because of a lack of metropolitan governments and a rather loose, 'distributed' system of municipal administration, the cities have little formal status in Australia's federal system. The Australian population is projected to grow by 48% (from 2016 level) over the next 30 years (Infrastructure Australia 2015) – putting pressure on where and how to accommodate this increase in the existing urban system. Australia is also very heavily dependent on fossil fuels for its energy and transport needs; however, state and some local governments have made strong commitments to transition to renewable energy. This context suggests that Australian cases can provide interesting test-tube perspectives on processes relevant to urban sustainability transitions worldwide.

Australia is also home to a growing community of transition researchers. Since early 2015, many of these researchers, scattered over the vast continent, have come together under the banner of ASTRA – the Australia-based Sustainability Transitions Researchers Alliance. This book, as you can see from the logo, is one of the early fruits of the collaborations in the ASTRA community. We will come back to the emergence and proliferation of Australia-based transition research in the concluding chapter.

This book is structured in four main parts. Part I, 'Introduction', contains three chapters providing further context for the Australian foci of the book and the ongoing conceptual and empirical challenges for understanding and pursuing urban transitions within the current transition framework and the notion of the urban age as one of planetary urbanisation. Part II, 'Governing Urban Transitions', includes four chapters relating how organisations and government policy are engaging (or not) with sustainability transitions to improve urban outcomes. Each chapter presents a different case study, respectively covering spatial planning (Chap. 4), organisational planning (Chap. 5), an extended interpretation of boundary organisations, looking at spatial as well as science-policy boundaries (Chap. 6) and exploring successful and unsuccessful government-supported niche development in Australia (Chap. 7). Part III, 'Specific Approaches to Urban Transitions', then explores urban sustainability transitions from the perspective of some of the key pillars for sustainable cities: water (Chap. 8), housing (Chap. 9) and decarbonising and relocalising the economy from below (Chap. 10). Part IV, 'Spatial Dimensions of Urban Transitions', includes three chapters which explore a historical review of housing policy development for creating cities (Chap. 11) and changes to how we use cities through live/work co-location (Chap. 12) and urban mobilities (Chap. 13). In the final section, Part V, a concluding chapter provides comment on the central questions and dimensions forwarded in this book and draws out key implications for transitioning to a low carbon, sustainable urban future.

Part I: Introduction

Chapter 1 by Horne builds upon the discussion and challenges of urban transitions presented throughout the previous volume (Loorbach et al. 2016). The chapter clearly identifies the requirement for a focus on urban transitions in the broader sustainability transitions field and how urban transitions research can inform the development of sustainability transitions theory and practice more broadly. It identifies continuing challenges for urban sustainability transitions. These include (1) multilevel governance challenges of transition management at the city scale where policy and regulation are invariably contested and conducted at multiple scales, (2) how to scale urban sustainability transitions in a post-neoliberal era of 'splintered urbanism', (3) how at the individual human interaction level spatial scales of transitions vary between cities and neighbourhoods and (4) how domestic-scale social practices interact with low carbon or sustainability transitions mechanisms. The chapter provides a detailed discussion of the different dynamics, challenges and mechanisms regarding purposive sustainability transitions within the urban context, building upon the insights articulated in the preceding book in the series. It also charts the benefits of mixed methods approaches in understanding transitions, including specifically the benefits of including household-level ethnographic investigations of social practices as ways of revealing how transitions in practice unfold in concert with transition mechanisms at the city scale. The Australian context for the book is further explored.

Building upon this, Chap. 2 by De Haan explores how concepts and frameworks used in the field of sustainability transitions do not have a spatial character and often do not address matters of place and scale at all. The chapter investigates challenges in pursuing purposive transitions in geographically defined settings working with sustainability transitions knowledge. De Haan argues that cities should be considered a nexus, as the collection of infrastructures and service provision systems, as well as a locus, the physical location and its sociohistorical identity where these systems reside or in relation to the broader systems (e.g. national or global) they are part of. A framework is proposed where urban transformative actors and the networks they form play a key role. Urban transitions processes are explained in terms of network dynamics with systemic consequences in the locus and the nexus aspect of cities.

In Chap. 3, Gleeson then explores the notion of the urban age as one of planetary urbanisation. Gleeson forwards that a two-stage transition is likely, the first stage being a move away from carbon capitalism followed by the second stage which is a new political economic order. Focusing on the first stage, Gleeson explores the possibility of a painful realignment to lead us to a more sustainable future and discusses that there may be a need for a strong 'guardian' state to guide such a transition, discussing such a transition using potential future scenarios.

Part II: Governance of Urban Transitions

Part II presents four chapters which look at different aspects of urban transitions governance. Morrissey et al. in Chap. 4 present a critical review of strategic spatial planning policy and outcomes at the city-region scale from Australia and Ireland since the early 2000s with a sociotechnical transitions framing. The chapter identifies a clear need for a new approach to strategic policy development with key sociotechnical transitions principles forwarded to reimagine and re-empower the practice of strategic spatial planning, better positioned to address future sustainability challenges. In Chap. 5, Bush et al. report on a case study investigating whether provision of targeted information on theories of sustainability transitions could strengthen organisational strategic planning and asks: if planning is informed by transition theories, would this assist and strengthen organisational visioning, ambition and confidence? The case study focuses on a community-based, not-for-profit organisation working on sustainable energy and climate change action in Melbourne, Australia. In addition to several key findings for applying sustainability transitions to organisations, the research demonstrates the potential impact of research-practice transdisciplinary partnerships in strategic planning but also continuing challenges around achieving this.

Following this, Moloney et al. (Chap. 6) discuss how regional 'boundary organisations' – those working across boundaries of multilevel governance and science-policy – might support innovation and enable transitions in public policy institutions and governance. The chapter reflects on insights gained from a collection of research projects working with different types of boundary organisations with interests in tackling climate change and promoting sustainability. The analysis shows how boundary organisations are contributing to improved local capacities and institutional shifts and outline factors enabling and limiting their potential in facilitating transitions towards sustainability under a changing climate. Rounding out this part of the book, Moore (Chap. 7) explores successful and unsuccessful niches in Australia through a strategic niche management framework. The chapter presents analysis and commentary of three recent government approaches to improving urban sustainability in Australia (two successful cases and one unsuccessful case): residential solar photovoltaics, ceiling insulation and exemplar high-density housing development. Each case contributes to a broader discussion about strategic niche management, including how the alignment with sociotechnical landscape dynamics leads to public sector actors having a greater chance at stimulating regime transformation.

Part III: Specific Approaches to Urban Transitions

This part of the book explores different sectorial perspectives on the application of transitions in the urban. In Chap. 8, Brown et al. look at the development over the past decade or so of an innovative approach that strives to establish water sensitive cities to address issues of water shortages caused by increasing droughts and population growth. The approach explores the integration of urban water management with urban planning, making extensive use of green infrastructure solutions. This chapter illustrates the uptake trajectories of elements of this approach, particularly under the severe stresses of the Australian Millennium Drought in Melbourne. Newton (Chap. 9) then looks at how large cities in Australia, like Sydney and Melbourne, traditionally try to accommodate population growth by growing larger in surface area – from their fringes outwards into the greenfields. The chapter discusses an alternative approach aimed at sustainable redevelopment of greyfield precincts through transition management and specifically the use of transition arenas.

Alexander and Rutherford (Chap. 10) conclude this part by exploring the Transition Towns movement (TTM) and the role such an approach could have on broader urban transitions. Although the TTM does not use transition theories explicitly in their approach, they arrive at similar conclusions. This chapter discusses what the TTM approach consists of and how that aligns with ideas about purposeful urban transitions. The chapter touches on the differences, challenges and opportunities for the TTM in Australia and how such an approach could be scaled up to challenge the dominant building and urban regime. The approach is illustrated by means of a utopian vision.

Part IV: Spatial Dimensions of Urban Transitions

In Chap. 11, Dalton explores the suburbs as a sociotechnical regime within Australian cities and explores the implications of the urban location of high-energy intensive household living within this regime. Dalton then discusses underlying 'lock-in mechanisms' producing and reproducing the suburbs and how this is destabilised and reconfiguring cities. The chapter concludes by drawing on a set of preconditions from the history of change in 'lock-in mechanisms' that should be considered in the development of transition to low carbon suburban suburbs and relates them to instruments in a policy framework. This is followed by Doyon (Chap. 12) who uses the example of live/work as a niche innovation within urban planning to develop a new framework for analysing the pathways of policy innovations within urban transitions. The research presents an embedded multiple-case study approach (Melbourne and Vancouver) with a theoretical replication design to explore contrasting results between different bounded systems. It is identified that rigid and top-down governance structures are less flexible and open to change, political approaches to planning are less responsive and adaptive and strong political actors have the ability to either initiate or inhibit change.

Concluding this part, Dodson et al. (Chap. 13) explore the equity challenges of urban sustainability transitions, particularly regarding sustainable transport in Australian cities. The chapter draws on empirical work by the authors revealing that the advantages of certain sustainable transport options, e.g. electric vehicles, are disproportionally more enjoyed by wealthier households. They find that a transition towards a highly fuel-efficient vehicle fleet will have adverse socio-economic consequences for lower-income households who typically live in outer suburbs and are, for that reason, highly car dependent. The authors suggest that systemic reform rather than market-led efficiency improvements are required to avoid such adverse equity outcomes.

Part V: Conclusions

The final chapter, Chap. 14, brings together the key contributions from the book and discusses their implications for future research and practice of urban sustainability transitions in Australia and internationally.

Trivess Moore
Fjalar de Haan
Ralph Horne
Brendan James Gleeson

References

Infrastructure Australia (2015) Population estimates and projections. Australian Infrastructure audit background paper April 2015. Canberra. Retrieved from http://infrastructureaustralia.gov.au/policy-publications/publications/files/Background-paper-on-demographic-projections.pdf
Loorbach D, Wittmayer JM, Shiroyama H, Fujino J, Mizuguchi S (eds) (2016) Governance of urban sustainability transitions. European and Asian experiences. Springer, Tokyo. https://doi.org/10.1007/978-4-431-55426-4

Contents

Part I
Introduction

Part I introduces the book and explores ongoing challenges for urban transitions from a practical, policy, and theoretical perspective. Chapter 1 locates the book into the emerging urban transitions discourse and why Australia presents an interesting focus for urban transitions research to inform international urban transitions. Chapter 2 builds upon this by analysing the conceptual issues of transitions studies in dealing with matters of place and scale. It brings into the discussion the distinction between essentially and accidentally place-based transitions and the notions of locus and nexus as a way of framing transitions when place matters. Concluding this section, Chap. 3 presents a provocative chapter about what a future urban transition may look like, in the face of failing neoliberalism. These chapters build the narrative around current and future challenges with urban transitions and set the context for the following sections in the book.

Chapter 1
Urban Low Carbon Transitions: Housing and Urban Change

Ralph Horne

Abstract The central question for this chapter is: *how can urban and transitions perspectives assist understandings of low carbon housing and urban change?* Current 'urban' and 'transitions' perspectives are presented along with recent and current attempts to bring urban and spatial perspectives to transitions studies. Australia as a site for urban transitions studies is considered, and three aspects of low carbon housing and urban change are highlighted: policy settings and governance, spatial/urban dimensions and carbon and consumption context. Contributions of urban and transitions perspectives to understanding low carbon housing and urban change are explored through two case examples of low carbon housing and urban change in Australia: photovoltaic panels on domestic rooftops and broader retrofitting and renovation activity towards low carbon housing. Transitions perspectives include the multilevel perspective and Transition Management. While these vary, the focus here is that they can each provide useful insights when coupled with other perspectives of urban and social change. Power, space and consumption all feature in practices of urban low carbon transitions, and it is essential that further analytical tools are brought to bear in these domains. They offer a scale for the study of cultural projects where change is as likely to be associated with cultural or social change as by policy settings.

Keywords Housing • Households • Low carbon • Photovoltaics • Retrofit

1.1 Introduction

In parallel to the twenty-first-century rush to cities, there is also increasing attention to urban theory and policy. More people are living in, focusing on and arguing about cities. The Intergovernmental Panel on Climate Change (IPCC) now specifically includes urban issues (IPCC 2014); 2015 saw the adoption of explicit 'urban'

R. Horne (✉)
College of Design and Social Context, RMIT University, Melbourne, VIC, Australia
e-mail: ralph.horne@rmit.edu.au

T. Moore et al. (eds.), *Urban Sustainability Transitions*, Theory and Practice of Urban Sustainability Transitions, DOI 10.1007/978-981-10-4792-3_1

imperative in the Sustainable Development Goals; 2016 saw the third bi-decadal UN Habitat conference and the 'New Urban Agenda', adding to a myriad of calls for actions in, on and for cities.

As introduced in above this book is primarily about the juxtaposition of the discipline of urban studies, loosely defined, with the relatively (as yet) aspatial ideas of sociotechnical transitions originating in science and technology studies. The rise of urbanisation and interest in cities is occurring contemporaneously with attempts to respond to climate change and the environmental sustainability crisis. In this chapter, some of the key points of both disciplinary traditions are introduced.

The central question for this chapter is: *how can urban and transitions perspectives assist understandings of low carbon housing and urban change?* This question is tackled first by introducing current 'urban' and 'transitions' perspectives, respectively, and the recent and current attempts to bring an urban or spatial perspective to transitions studies. Then Australia as a site for urban transitions studies is considered, followed by the main section of the chapter where three aspects of low carbon housing and urban change are highlighted, as follows:

- Policy settings and governance
- Spatial/urban dimensions
- Carbon and consumption context

In highlighting these aspects, of course, others are diminished. These three are selected specifically in order to draw out contributions of urban and transitions perspectives to understanding low carbon housing and urban change. Two case examples of low carbon housing and urban change in Australia are then presented in order to illustrate how urban and transitions perspectives can contribute in different ways to our understanding of this phenomenon.

By transitions, I refer to purposive, sustained sociotechnical interventions in existing systems or 'regimes' (Geels 2002, 2010) designed to shift them out of the way or to transform them – in this case, leading to a new, 'sustainable' low carbon setting. There are significant variations between transitions perspectives that are outside the scope of this chapter. 'Transition Management' (Loorbach 2007) involves setting a specific agenda involving the bringing together of decision makers (both policy and business entrepreneurs) to set collective visions and strategies and to act on these. Elsewhere, the multilevel perspective (MLP) combines ideas from science and technology studies as well as policy and management, evolutionary economics and the sociology of innovation in a heuristic schema for understanding sociotechnical change (Geels et al. 2016).

The rapidly growing literature extends well beyond the expected urban/geography/planning and technology/policy/management traditions – themselves rich, diverse and dynamic. Concepts and methods clash and bump alongside and across each other in a sea of concern to understand and articulate meanings to concepts such as 'prospects', 'purposive' or 'transitions'. Scholars of design, engineering, politics, sociology, anthropology and beyond have been drawn to comment on a set of overlapping topics, from questions of governance, sustainability and resilience to considerations of (to name a few) boundaries, practice, agency, materiality and culture.

1.2 Urban Transitions in Australia

Urban studies itself spans a range of disciplines and is expanding and in some aspects diversifying, just as cities themselves are. For some, the urban project has been celebrated as a competitive 'triumph' (Glaeser 2011; Brugmann 2009, etc.). Cities are being variously labelled smart, resilient, sustainable and so on, yet this technological optimism is set against a backdrop of rapidly worsening inequality, resource scarcity and climate change. Although the rhetoric on social and policy change has grown over recent years, overwhelmingly the focus of urban decarbonisation plans is technology substitution (e.g. IPCC 2014). It seems market-based promotion of renewables and resource efficiency remains the great hope for cities. Such hope is set against that fact that much of the urban sustainability equation – such as energy grids – lie outside cities' direct control.

Cities are more than material inventions; and they are more than clusters of people who choose to live in close proximity. They are characterised by geographic differentiation, where proximity is a key organising theme – what Storper and Scott (2016) call the *urban land nexus*. While there is no doubt that cities – in both the physical and social sense – shape us as much as we shape them, current urbanisation patterns are also clearly and increasingly obviously reflections of political–economic practices of globalisation and capital accumulation. Unsurprisingly, many urban scholars (e.g. Gleeson 2014) are therefore concerned with the extent to which cities can intervene in the neoliberal project or whether they are ultimately subordinate to it.

In this chapter, the starting point is the idea that cities are about people and the policy of proximity. This plays out in daily patterns of commutes, work and spatial–temporal rhythms. Cities can be seen as both locus and nexus (see Chap. 2), i.e. as a number of co-located elements or as a more coherent organism entity. Similarly, we can differentiate between phenomena that unfold *in* the city and those that help define and therefore are *of* the city. For example, the fact that photovoltaic (PV) panels start appearing on city rooftops does not mean they are *of* the city in an urban sense – a topic we return to later in this chapter.

Contemporary urban theory debates that are beyond the scope of this chapter variously advance post-colonial urbanism (e.g. Roy 2011; Robinson 2011), assemblage theory (e.g. Farías and Bender 2010; Marston et al. 2005) and planetary urbanisation (e.g. Lefebvre 1970; Brenner and Schmid 2015). Notwithstanding the implications for urban studies, whether about boundaries, hierarchies or distinctiveness arising from colonial or other historical experiences, the starting point for our investigation of cities facing urban low carbon transitions is that human agency and urban place theory matter. At the same time, ideas of distributed agency are also critical. No doubt, the complexities of urban place making and remaking can be understood through deep, detailed investigations of social practice, materially imbricated urban lives and ethnographic inspired explorations of the urban web of life. Thus, reflexive relations between technology, urban space, social life and political–economic forces are critical in understanding the urban (Guy et al. 2001).

1.2.1 Transitions and the Urban Turn

The origins of Transition Management point to an original understanding of the place of technology in society. The ontology is one of industry changes that relies upon introduction of adoption of technologies in novel ways and the imbrication of this innovation with social change, new knowledge about 'how' and new policy settings that, in some way or another, favour the emergences of the novel and the new to such an extent that, over time, the existing status quo is displaced or replaced or otherwise disappears in favour of a new regime.

Technology transitions require a set of institutional and social factors to be in place and often struggle through a period.

The MLP thus seeks to offer a simple explanation of dynamics of system obduracy and change of how seemingly 'set' regimes can be usurped and disappear. It has become a popular heuristic for explicitly separating out multiple 'levels' (regime, niche, landscape) to be accounted for in terms of policy, power and institutional alignments. The issue of governance is at the heart of the idea of purposive transitions; it is the very levers of policy and governance that are advocated as central features in the engine room of the transition.

This is not to overplay the role of government and certainty of success if such tools. As so many actors are involved and interests inevitably conflict and struggles ensue with unpredictable outcomes, 'system innovations are characterized by emergent and nonlinear dynamics' (Geels et al. 2016:2; O'Brien 2015).

This emergent yet growing genre of sociotechnical transitions research has been largely aspatial and silent on questions of cities, until recently (e.g. Coenen et al. 2012). As introduced in the foreword to this book, Transition Management is now engaging with the urban context (Loorbach et al. 2016; especially Chap. 1) and there is a rapidly growing regional and urban Transition Management and 'geography of transitions' literature (Hansen and Coenen 2015). This spatial turn parallels a longer-held concern by urban scholars about prospects for low carbon cities; notable contributions that consider proximity and geography of cities in this context include UK scholars Bulkeley, Guy and Marvin. We return to questions of how transitions might manifest in cities (or how cities might themselves transition) below.

1.3 Why Australian Cities?

We advance three principal reasons for considering Australian cities as sites for urban transitions studies. Firstly, as introduced in above this book is the first collection of research and practice representing recent developments in the urban transitions field, broadly defined, in the Australian region and has merit on this basis. Second, there is the prospect of patterns of practice emerging that are regionally specific. In other words, it is prima facie logical to collect geographically proximal studies that involve specific spatial settings as we can expect links between them – a topic we return to in the concluding chapter in this book. Thirdly, and more speculatively, Australian urban environments provide a focus for transitions researchers and practitioners and an international audience because of the particularities of their location, form and political, social and cultural context.

Australian cities are relatively spread apart geographically, so they offer reasonably contained ‘living laboratories’. They are low density, new world, westernised, yet geographically remote from their closest cousins in North America and Europe. Urbanisation is relatively advanced – Australia is one of the world’s most urbanised populations with more than 82% of the population living in cities. Moreover, Australia has been economically heavily fossil fuel dependent, with very high per capita emissions and an economy heavily based upon mining, including fossil fuels. It also has a complex political system in which climate change itself is contested at the federal level, while at the local level (at least in principal cities), it is generally understood and there are many hundreds of initiatives amongst local authorities and communities designed to adapt and mitigate climate change.

Yet, despite this complexity and contestation, there is some evidence that social and cultural settings are amenable to rapid shifts in uptake of new low carbon technologies and practices. Indeed, the domestic retrofit of solar-PV systems, from a few thousand in 2008 to some 1.5 million in 2016, is one of the most rapid ‘transitions’ of its type, and it continues apace across the 7.5 million detached dwellings in Australia, despite the wholesale removal of economic incentives and the increasing imposition of disincentives (see below and also Chap. 7 by Moore).

1.4 Three Questions

1.4.1 Policy Settings: Who Is Steering and for Whom?

In line with ideas of ecological modernisation (Mol and Spaargaren 2000), Transition Management holds that low carbon transitions can be delivered by capitalism, and, in advocating for this, it has tended to give primacy to the levers of government or, at least to large institutions, in instigating and convening transitions. Apart from questions about the uncertain future of capitalism and the capacity of the neoliberal project to deliver such change, the idea of Transition Management has been problematised, not least on the grounds that ‘regimes constitute the selection environments in which niche innovations fail or flourish, and which emphasise processes of alignment and path dependence’ (Shove and Walker 2010:472). Notwithstanding that Transition Management espouses the laudable aim of building coalitions of willing policy and business entrepreneurs and empowering civil society around shifting normal practice to low carbon alternatives, there is the problem that established interests and power relations are at stake. Moreover, ideas of co-design of long-term transitions can risk naivety, as revealed in a study of the micropolitics of transitioning in the Dutch transport sector (Avelino 2009).

Through detailed empirical work, Avelino investigates the 'ironic situation in which policies that are designed to 'empower' people, in themselves require people to already be 'empowered'' (ibid:369). Decision makers, business and government, tend to dominate the visioning and plan towards sustainability, while the weaker segments of society, who can be said to suffer the most from 'unsustainability', are excluded either explicitly or through the discourse, language and mode of engagement adopted. Thus, 'in an era of 'sustainability' discourse, the first P of the people–planet–profit–balance is often forgotten as discussions submerge in quantitative squabbling over cost–benefit analysis and trade-offs between ecological and economic targets' (ibid:388).

Although from different traditions and persuasions, both Geels (2010) and Shove and Walker (2010) would surely agree that the thorny question of governing of transitions and/or social practices needs attention, particularly the role of civil society and social movements and the multiple agents (human or nonhuman) involved (indirectly or directly) in the replication of 'high carbon' or the introduction of 'low carbon' ways in cities. While this debate has value in itself, for current purposes, the question arises about how considerations of power and governance confront low carbon transitions in/of cities.

Perhaps most obviously, when it comes to governance and power in cities, there are invariably considerations of multilevel governments and governance. Cities may have mayors with local powers, but invariably there are one or two other tiers of government above and perhaps even below, who also share or have other powers. This said, cities can operate directly with international institutional relationships, such as in the Cities for Climate Protection programme, bypassing and potentially progressing pro-climate action against the policy of the nation state, as in the case of Australia and the USA (Bulkeley and Betsill 2003). The spatial specificity of cities (with apologies to Lefevre and Latour) help illuminate the loci of power in transitions. This illumination, it must be said, is still a work in progress. As Castan Broto and Bulkeley (2013) point out, methods of data collection about low carbon experiments in cities tend to favour those with local capabilities and resources, which tend to be institutions in richer (northern/western) cities. Hence, it has largely overlooked the responses emerging outside formal contexts of decision-making and led by actors other than municipal governments and also initiatives in southern/developing cities.

While there are distinctions between them, a range of literatures highlight the importance of (a) networks in governing low carbon action and (b) the uncertainty of outcomes of such actions. This gives rise to the idea of low carbon experiments towards transition (Moloney and Horne 2015) drawing on governance experiments (Hoffman 2011), strategic niche management and grassroots innovations in sociotechnical regimes (Geels et al. 2011) and ideas of 'urban laboratories' (Evans and Karvonen 2014) in sociotechnical change (Castan Broto and Bulkeley 2013). Low carbon urban experiments might also be thought of at the whole-of-urban scale, such as in Fastenrath and Braun's study of Freiburg (Fastenrath and Braun

2016). Such experiments can be led by city leaders, power brokers, institutions and governments, but they can also be led by civil society groups who are otherwise on the outside of city low carbon policymaking.

1.4.2 Spatial/Urban Dimensions: Do Cities Transition or Do Transitions Inhabit Cities?

As introduced above, the spatial turn is underway in sociotechnical transitions studies, and there is now an emerging literature on the geography of transitions. This attention to space emphasises diversity in, for example, urban low carbon transition processes arising from a 'natural' variety in institutional conditions, networks, actor strategies and resources across space (Coenen et al. 2012). It holds the possibility of being able to a say *where* low carbon urban transitions might be, rather than describing them on more technical, abstract and aspatial terms. It creates space to consider place-specific relationships, such as the role of local interpersonal contact in understanding the co-evolution of knowledge, practice and institutional engagement. Here, we are reminded of the importance of proximity in general theories of capitalist production and consumption, as well as in urban studies (c.f. Storper and Scott 2016, discussion above).

While the MLP advances ideas of niches and the build-up of 'local' actors who then become 'global' in their outlook, this shift is seen as cognitive rather than spatial – yet, geography is surely important. What does the progress of a transition or niche look like in cities? Do we expect larger clusters of experiments or do they take on a different form? Van Doren et al. (Van Doren et al. 2016) differentiate between horizontal scaling up initiatives where 'diffusion' (Rogers 1995) occurs as a number of projects or initiatives replicate and spread out and vertical scaling up, where concepts, ideas, values and knowledge spread upwards (implying a vertical structure of governance) and/or otherwise become institutionalised, embedded in the mainstream, etc. (Bai et al. 2010). In any event, what Coenen et al. (2012) call 'inter-localisation' is actively constructed by low carbon innovators through socio-spatial struggles. Following those relationships and struggles reveals ways within which experiments might interrelate and 'upscale', in turn allowing improved understanding of relations between these experiments and the prospects for broader sociotechnical shifts.

While a spatial lens is a useful start, it does not address the question of whether (or to what extent), say, niche experiments might inhabit a city and gradually populate it spatially or the extent to which a city might act as an entity. This can be summarised as the 'city transition as locus' vs 'city transition as nexus' discussion, which will be taken up in more detail in Chap. 2. In the meantime, it is a valid question, we argue, for all studies of low carbon urban actions and experiments that are concerned with the prospects for wider urban transitions.

1.4.3 Carbon and Consumption Context: Can a City Remove Its Own Bedrock?

While this question is to some extent entangled with the two previous questions, it serves to emphasise further the challenges of low carbon governance. Carbon is urban bedrock; it is fossil fuels that have fuelled – literally and figuratively – the modern era of urbanisation. This is not to proclaim an inelastic or causal relationship between carbon emissions and urban growth, but rather to point out that the relationship has run deep, at least up to the present. The era of capital accumulation and rush to cities has been highly dependent upon an economic system of exchange where the environmental costs of fossil carbon-related (i.e. almost all) goods and services are 'externalised' or otherwise ignored. Cities have been built on the convenience of fossil fuels that have been extracted and utilised in such a way that the long-term global costs of doing so have been ignored. Cities are as such enabled through a tacitly agreed global environmental debt.

In this context, how does Transition Management translate to the specific, but somehow dispersed, idea of urban low carbon transitions, in cities where carbon consumption is quite literally 'built in' through combustion engine-based systems of private transport and through fossil-based grids of energy reticulation? While this question applies to some extent to transition studies more broadly than just in the context of the urban and to cities more broadly, there are particular reasons why the 'bedrock' concept applies to Australian cities (Horne and Fudge 2014). First, Australian cities are particularly high emission, low efficiency (in carbon per capita terms) by global standards. This has the effect of widening the gap between current trajectory and expected future trajectories. Second, the national economy is resolutely primary industry based, reliant upon cheap exports of fossil fuels. This has profound implications for national policy priorities and for the wealth of cities, which is based significantly upon providing financial and related services that are directly linked to fossil fuel extraction and export. Third, a succession of state governments since 2010, and federally since 2013, has made shifts of policy language away from climate change mitigation. The shifts range from outright climate change denial to a nuancing of policy settings towards adaptation or cost–benefit-based energy efficiency.

Of course, none of these settings are permanent, and there are a range of more progressive trends reconfiguring the economy, the policy rhetoric and the materiality and cultural practices of urban communities, as we shall see in the following examples. Moreover, in MLP language, there are landscape factors at play, with COP21 and the ongoing efforts towards international binding policy galvanisation around a 450 ppm CO_2 concentration maximum. In this scenario, global demand for coal will fall by 30% over the next two decades, giving way to a mix of energy efficiency and renewables. This scenario illustrates the exposure of the current Australian economy and urban configurations and perhaps explains why Australian cities are actively engaged in urban low carbon experiments, both locally and connected to international networks.

1.5 Domestic Perspectives: Two Case Examples

1.5.1 The Great Australian Domestic PV Transition

In striking contrast with its high fossil fuel dependency, Australia has managed a rapid shift in domestic PV from practically zero systems in 2007 to now (2016) having the highest proportion of households with PV systems on their roof of any country in the world. Over 15% of Australian households have solar panels on their roofs (some 1.5 million systems).

Initially, this may appear as a simple case of ecological modernisation in action using the tools of Transition Management. The 'rational' potential for domestic PV was apparent (Sivaraman and Horne 2013), and a set of federal and state market mechanisms were instituted to support the transition, with the result that a technical substitution was achieved, namely, the replacement of fossil-fuelled grid capacity with a distributed PV system owned by millions of householders.

However, there is more to the story. Firstly, the federal support via a generous Renewable Energy Certificates scheme was at least in part a rapid Keynesian response to the global financial crisis in 2008–2009, rather than a considered long-term strategy. Moreover, the rapid shift created genuine concern on the part of the private monopolies and large companies who have significant sunk assets in grid infrastructure. Following lobbying and a change of government, the subsidies were rolled back. States reigned in generous feed-in tariff arrangements, from 68c per kWh to 6c per kWh, against a typical fossil energy retail price of 25c per kWh. The closures and rollback of market mechanisms took place between May 2011 and September 2013. After this date, it was uneconomic under any 'rational' potential scenario for householders to invest. Yet, as shown in Fig. 1.1, PV installations have continued to the present with only minor ripples as the various schemes ended.

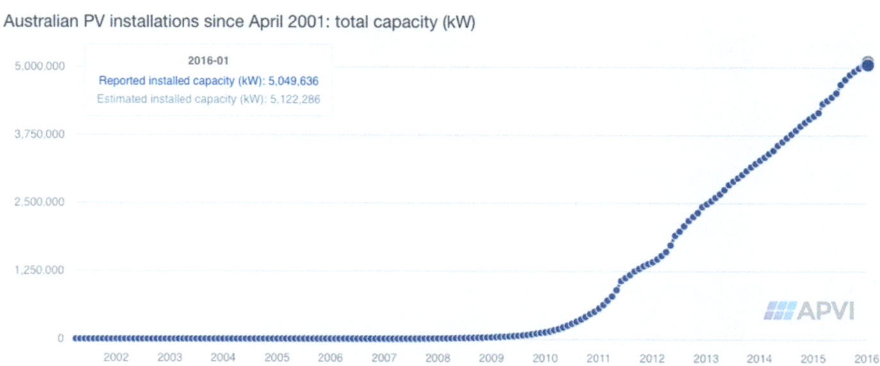

Fig. 1.1 Solar PV installations in Australia since 2001 (Source: Australian PV Institute: http://pv-map.apvi.org.au/analyses)

Spatially, 40% of solar installs were in rural and regional communities (up to 2014), despite these communities making up only 32% of Australia's housing stock. There are significant variations between states, but common factors amongst areas with high concentrations were that they tended to be areas with high concentrations of older, low-income households; areas with high concentrations of detached, owner-occupied housing; and areas with high levels of new home-building or renovation activity (Green Energy Trading 2014). Nevertheless, these spatial (concentrations of new buildings, existing detached buildings and areas of renovation activity) and characteristics help explain the 'how, where and who' of the unfolding domestic PV transition. This is consistent with PV early market formation in Germany, where localised constituencies were at play, whereas 'An a-spatial analysis of success factors would easily conclude that regulatory incentive structures have been the sole relevant factor and overlook the local formation processes which built the economic, political and instrumental basis which enabled these regulations to gain widespread political acceptance' (Coenen et al. 2012:970).

Reflecting on these two observations, we speculate that 'rational choice' is not the dominant factor at play and that the market intervention served to kick-start a cultural project, during which the idea of cost-effectiveness of PV was established. Hence, it took root in households concerned about high energy bills as well as those concerned about carbon. This idea then became part of a broader narrative about PV systems being the 'new normal' essential housing add-on, which has allowed installations to continue despite the abrupt end to policy settings that provided cost incentives. Again, this is commensurate with MLP-type framing of transitions as socially and culturally imbricated projects and perhaps explains why in other parts of the world also 'the diffusion of solar-PV, onshore wind turbines, and LED-lighting has been faster in recent years than was anticipated' (Geels et al. 2016:4).

Also of interest are the potential unintended consequences of this market-based policy experiment. What are the disruptive prospects for an incumbent set of actors with a business model predicated upon continuation of a supply–demand divide, now that millions of householders may soon have access to storage technology to manage 'their' new PV resource without the need for a grid connection? This question is all the more pertinent for the fact that it has proved difficult to 'stop' PV installations. By the time regime actors attempted to do this, the transition had converted into a sociocultural project (see also Chap. 7 for further evaluation of the domestic PV development in Australia).

1.5.2 Low Carbon Housing Retrofit and Renovation?

In contrast to the PV example, the characteristics of a low carbon housing retrofit and renovation transition present uncertain technological and definitional boundaries. Also, once we climb off the rooftop and enter the house (as many retrofit and renovation technologies require), a broader set of concerns around the home are encountered (Horne et al. 2014). Although the boundary is somewhat porous, by retrofit, I refer to technology substitution or add-ons, such as draft-proofing, double glazing, eco-appliances, etc., whereas renovation invariably involves a wider range of structural and cosmetic modernisations and updating, in concert often with retrofit. If we take the scenario where householders are primary actors in taking decisions to undertake or commission low carbon retrofit (or to lobby their landlords to commission such work), then there is the obvious potential clash between concerns for carbon, on the one hand, and other concerns about 'comfort, cleanliness and convenience' (Shove 2003), on the other.

Low carbon housing retrofit and renovation often entails material and cultural entanglements with the intimacies of domestic life. While some retrofit technologies, such as insulation, lurk behind walls and in ceiling spaces and are rarely directly confronted by households and domestic practices, others and especially those combined with renovation tend to be more intrusive. Thus, low carbon retrofit and renovation are immediately caught up in intersections of daily routines, broader housing aspirations, narratives of housing sustainability (Maller et al. 2012) and ideas of what constitutes 'good' sustainability citizenship (Horne et al. 2016).

The dominant mode of low carbon retrofit in homeowner dominated and detached dwellings across Australian cities; householders negotiate with builders and 'eco-retrofitters' and respond to government incentives and a myriad of commercial and civil society messages about notions of the 'good home(-maker)', the 'green home', etc. The technologies vary with the scope of retrofit. Research investigating the emergent niche around Melbourne, Australia, found two main types of activity (Horne et al. 2014). The first was smaller-scale retrofit, ranging from LED light bulbs and low flow showerhead exchanges, through weather stripping and roof insulation, to add-on rainwater tanks and solar systems. The second was larger-scale renovation, ranging from kitchen and bathroom replacements to structural alterations or additions. Accordingly, two types of service providers are involved at the front line: specialist 'eco-retrofitters' who generally work with a limited range of ecotechnology add-ons and project managers (eco-builders, eco-designers) who specialise in trying to create low carbon housing renovations.

To the extent that there might be a discernable niche, the boundary between this and the considerable 'mainstream' home improvements industry is blurred. While householders might highlight their projects as either 'eco' or 'mainstream', some project managers work in both, and elements of what might be called 'eco' and 'mainstream' home improvements span a wide range of projects. Indeed, 'The only difference that could be detected is that green renovators explicitly overlay sustainability to an already complex set of explanations, and this was associated with a higher level of awareness and intent to find out and test 'green' ideas' (Horne et al. 2014:22). In other words, the sustainability narrative of 'mainstream' renovators was more muted, but crossover technologies and elements were found in both types of projects.

One way to indicate a potential low carbon retrofit niche boundary is through lines of contestation. Households push project managers to adopt low carbon innovations in some cases, often resulting in negotiated mutual co-self-education and experimentation; and, in other cases, project managers push back with householders who are seeking to prioritise comfort or other ‘non-eco’ changes. Also, these struggles occur in conjunction with other discernable ‘niche-like’ attributes. For example, changes in new housing regulations enable product substitutions (e.g. double glazing and solar hot water systems become more cost-effective), and these in turn find their way into housing renovations as these products become more ubiquitous. Government rebate and other market mechanisms of support are also key, especially in some smaller-scale retrofit businesses (eco-product exchanges, weather stripping and roof insulation as well as solar technologies have benefitted here). Hence, the stop–start and varied quality of these schemes is a problem for an emergent low carbon retrofit niche. Moreover, niche-type renovation interventions are imbricated with household preferences, mediated by a wide range of intermediaries (Horne and Dalton 2014).

There are no detailed studies of the spatial dimensions of low carbon housing retrofit in Australia, but we know from interviews with low carbon retrofit businesses that they are overwhelmingly small, they operate locally, they typically find work within areas of high ‘mainstream’ renovation (e.g. older, owner-occupied, detached housing stock) and they operate within heuristic, localised, formative peer networks. Their backgrounds vary, with small-scale retrofitters especially likely to have had different previous careers (IT, public service, etc.). They may have worked in different cities and different parts of the construction industry, but in general they now follow referrals as they build their business; this invariably means being geographically focused in one part of the city, but following jobs beyond this across the entire city as required.

A combination of the primacy of trust and the fact that the work is generally bespoke, taking place in people’s homes, means that low carbon housing retrofit industries necessarily have spatially localised characteristics. Built upon relationships, much of the work involves some form of co-production and negotiation. This work, imbricated with the intimacies of domestic life, is a far cry from IPCC or Government messages about carbon reduction targets (Karvonen 2013). It suggests that low carbon housing retrofit might take on particular characteristics in particular cities, just as other aspects of cultural life do. It is part of multiple bundles of low carbon transitioning (mobilities, work, leisure, etc.) and itself could be regarded as multi-faceted; here, we have only articulated two main types, but further interrogation would undoubtedly reveal more.

1.6 Discussion and Conclusion

So, *how can urban and transitions perspectives assist understandings of low carbon housing and urban change*?

Table 1.1 summarises key aspects of the two case studies with regard to the three factors posed above, namely, policy settings, spatial/urban dimensions and carbon characteristics. The domestic PV transition in Australia can to some extent be usefully understood as a sociotechnical transition in MLP terms. It involves

Table 1.1 Summary of key aspects of the two case studies

	Policy settings	Spatial/urban dimensions	Carbon and consumption context
Domestic PV	At first, federal and state incentives	More demographic than urban?	Low carbon technology substitution, social structuring
	Now, a cultural project?		
Retrofit and renovation	Haphazard and contested governance	Both locus and nexus? Complex bundles of urban consumption practices	Diverse technologies. Assumed or deemed substitution. Carbon and renovation both imbricated with wider social structures of consumption
	Some evidence of niche activities	Bundles of low carbon STT experiments (PV, insulation, double glazing, etc.) entwined with the broader urban renovation regime	

only a limited range of technologies and practices and an identifiable range of ‘steering’ supporting measures. It holds out the prospect of low carbon technology substitution, although at least two other possibilities exist; that having PV leads somehow to increased energy use; or that it leads to reduced energy use. There is some limited suggestion that some people reliant on localised but grid-connected energy systems are more conscious of how and when they use electricity and have a lower energy use compared to people solely reliant on the mains grid (Hondo and Baba 2010). An urban perspective, on the other hand, brings more questions about how socio-spatial dynamics unfold, the location and mobility of providers of services, the economic and demographic and tenure characteristics of households and the structuring of PV as a potential ‘new normal’ of socio-material relations.

Similar points can be made about low carbon urban housing retrofit more broadly. This, of course, is more difficult to define and has elements of both ‘locus’ and ‘nexus’. It is more closely imbricated with broader ‘mainstream’ renovation practice, much of which is expressive of broader patters of consumption, including changing aspirations and expectations, for example, of comfort. Carbon and renovation are both imbricated with wider social structures of consumption. Renovation might ostensibly include ‘green’ technologies or design and/or may also enable new modes of (potentially) elevated consumption, for example, through extensions and additions or multimedia spaces. In MLP terms, there are niche-like aspects to retrofit and renovation, variously defined, but there are multiple boundary problems and a wide variety of changing policy settings, diverse service providers and, of course, considerable socio-spatial heterogeneity.

Both cases illustrate the utility of the MLP heuristic or middle range theory in introducing ideas of niches, regimes and landscapes. It is also limited and, possibly, limiting. It offers little spatial or urban sensitivity or means to connect the ‘social’ and ‘technical’ and the numerous agents (both human and nonhuman) involved in maintaining and destabilising or reconfiguring regimes. Most focus in transitions studies has been on institutions and technologies rather than the spaces between them. Exceptions include studies of intermediaries, in-betweens who intentionally work to advocate, network, innovate and variously intervene in existing modes of doing things by influencing multiple actors (Horne and Dalton 2014; Backhaus 2010; Moss 2009; Fischer and Guy 2009; Guy et al. 2011; van Lente et al. 2003). Intermediaries can be seen as more horizontal, bottom-up change agents; 'the character and work of intermediaries are defined by their 'in-between-ness', rather than their organisational structure or any particular focus of their work' (Moss 2009).

This offers a different emphasis to Transition Management, where a key emphasis is placed on the levers of regulation, policy and government. The study of transactors and intermediaries offers fertile ground for extending transitions ideas into the urban. A concrete example is the niche 'green' policy/business/social enterprise/civil society entrepreneur, working across the city from neighbourhood to neighbourhood and from builders to community groups to city hall. Intermediaries offer the potential to gather heuristic skills, relationships and trust along the way, providing both 'glue' for assembling coalitions of the willing and a deep, nuanced understanding of social and cultural dimensions at play. In turn, operating at the local and city scale, they 'could potentially support the design and delivery of programmes that more adequately recognise the modus operandi of the industry, households and civil society organisations' (Horne and Dalton 2014:3457). Insights from detailed, fine-grained, practice-based research has benefits here as 'it emphasizes the importance of broad coalitions of actors (for example, project developers, local authorities, citizens, local shop owners and community groups) . . . as participants in innovation projects ..[involving] . . . trust, cooperation, commitment and collective action' (Geels et al. 2016:5).

In an era of individualised, often balkanised, private interests, the task of purposive urban low carbon transitioning must at minimum involve nurturing coalitions of the willing in dynamic, evolving, spatially situated bundles of practices, comprising materials and technologies, skills and knowledge, policy settings and social rules. Ideas of niches within the MLP and levers of governance in TM can be a useful starting point but do not go far enough in revealing the role of power, space and consumption. This is not to argue for alternatives; rather, both urban and transition perspectives are useful for different purposes (Geels et al. 2016) and for understanding urban change at different scales. Rather than rollout of a blueprint, low carbon urban experiments are part of emergent, learning-by-doing processes (Bulkeley et al. 2014). Australian cities provide particular opportunities and points of intervention; they can shed new light on urban transitions studies more broadly. They offer a scale for the study of cultural projects where change is as likely to be associated with cultural or social change as by policy settings.

References

Avelino F (2009) Empowerment and the challenge of applying transition management to ongoing projects. Policy Sci 42(4):369–390

Backhaus J (2010) Intermediaries as innovating actors in the transition to a sustainable energy system. Cent Eur J Public Policy 4:86–108

Bai X, Roberts B, Chen J (2010) Urban sustainability experiments in Asia: patterns and pathways. Environ Sci Pol 13:312–325

Brenner N, Schmid C (2015) Towards a new epistemology of the urban. City 19:151–182

Brugmann J (2009) Welcome to the urban revolution. Bloomsbury Press, New York

Bulkeley H, Betsill M (2003) Cities and climate change: urban sustainability and global environmental governance. Routledge, London

Bulkeley HA, Broto VC, Edwards GAS (2014) An urban politics of climate change: experimentation and the governing of socio-technical transitions. Routledge, Oxfordshire/New York

Castan Broto V, Bulkeley H (2013) A survey of urban climate change experiments in 100 cities. Glob Environ Chang 23(1):92–102

Coenen L, Benneworth P, Truffer B (2012) Toward a spatial perspective on sustainability transitions. Res Policy 41(6):968–979. Available at: http://dx.doi.org/10.1016/j.respol.2012.02.014

Doren DV, Driessen PPJ, Giezen M (2016) Scaling-up low-carbon urban initiatives: towards a better understanding. Urban Stud. doi:10.1177/0042098016640456

Evans J, Karvonen A (2014) 'Give me a laboratory and I will lower your carbon footprint!' – urban laboratories and the governance of low carbon futures. Int J Urban Reg Res 38:413–430

Farías I, Bender T (2010) Urban assemblages: how actor-network theory changes urban theory. Routledge, London

Fastenrath S, Braun B (2016) (In Press) Sustainability transition pathways in the building sector: energy-efficient building in Freiburg (Germany). Appl Geogr 10:1–11

Fischer J, Guy S (2009) Re-interpreting regulations: architects as intermediaries for low-carbon buildings. *Urban Stud* 46:2577–2594

Geels F (2002) Technological transitions as evolutionary configuration processes: a multi-level perspective and a case study. Res Policy 31:1257–1274

Geels F (2010) Ontologies, socio-technical transitions (to sustainability), and the multi-level perspective. Res Policy 39:495–510

Geels FW, Berkhout F, van Vuuren DP (2016) Bridging analytical approaches for low-carbon transitions. Nat Clim Chang 6:576–583

Glaeser E (2011) The triumph of the city: how our greatest invention makes us richer, smarter, greener, healthier, and happier. Penguin, Harmondsworth

Gleeson B (2014) The urban condition. Routledge, London/New York

Green Energy Trading (2014) Postcode and income distribution of solar. http://www.recagents.asn.au/wp-content/uploads/2014/04/GET-Postcode-report-for-RAA-April-2014.pdf

Guy S, Marvin S, Moss T (2001) Urban infrastructure in transition. In: Networks, buildings, plans. Earthscan, London

Guy S, Marvin S, Medd W, Moss T (2011) Shaping urban infrastructures: intermediaries and the governance of socio-technical networks. Earthscan, London

Hansen T, Coenen L (2015) The geography of sustainability transitions: review, synthesis and reflections on an emergent research field. Environ Innov Soc Trans 17:92–109

Hoffman M (2011) Climate governance at the crossroads: experimenting with a global response after Kyoto. Oxford University Press, Oxford

Hondo H, Baba K (2010) Socio-psychological impacts of the introduction of energy technologies: change in environmental behavior of households with photovoltaic systems. Appl Energy 87:229–235

Horne R, Dalton T (2014) Transition to low carbon? An analysis of socio-technical change in housing renovation. Urban Stud 51(16):3445–3458

Horne R, Fudge C (2014) Low carbon urban Australia in a time of transition. In: Miller C, Orchard L (eds) Australian public policy progressive ideas in the neo-liberal ascendency. Policy Press, Bristol, pp 279–295
Horne R, Maller C, Dalton T (2014) Low-carbon, water-efficient house retrofits: an emergent niche? Buil Res Inf 32(1):17–32
Horne R, Fien J, Beza B, Nelson A (2016) Sustainability citizenship: living and working sustainably in our cities. Routledge, London
IPCC (2014) Climate change 2014, Synthesis Report. Intergovernmental Panel on Climate Change
Karvonen A (2013) Towards systemic domestic retrofit: a social practices approach. Build Res Inf 41:563–574
Lefebvre H (1970) La revolution urbaine. Gallimard, Paris
Loorbach D (2007) Transition management: new mode of governance for sustainable development. Erasmus University Rotterdam. International Books, Utrecht
Loorbach D, Wittmayer JM, Shiroyama H, Fujino J, Mizuguchi S (2016) Governance of urban sustainability transitions: European and Asian experiences. Springer Japan, Tokyo
Maller C, Horne R, Dalton T (2012) Green renovations: intersections of daily routines, housing aspirations and narratives of environmental sustainability. Housing Theory Soc 29(3):255–275
Marston SA, Jones JP, Woodward K (2005) Human geography without scale. Trans Inst Br Geogr 30:416–432
Mol A, Spaargaren G (2000) Ecological modernisation theory in debate: a review. Environ Polit 9:17–49
Moloney S, Horne R (2015) Low carbon urban transitioning: from local experimentation to urban transformation? Sustainability 7:2437–2453
Moss T (2009) Intermediaries and the governance of sociotechnical networks in transition. Environ Plan A 41:1480–1495
O'Brien K (2015) Political agency: the key to tackling climate change. Science 350:1170–1171
Robinson J (2011) Cities in a world of cities: the comparative gesture. Int J Urban Reg Res 35:1–23
Rogers EM (1995) Diffusion of innovation. The Free Press, New York
Roy A (2011) Slumdog cities: rethinking subaltern urbanism. Int J Urban Reg Res 35:223–238
Shove E (2003) Comfort, cleanliness and convenience: the social organization of normality. Berg Publishers, Oxford
Shove E, Walker G (2010) Governing transitions in the sustainability of everyday life. Res Policy 39(4):471–476
Storper M, Scott AJ (2016) Current debates in urban theory: a critical assessment. Urban Stud 53(6):1114–1136
van Lente H, Hekkert M, Smits R, van Waveren B (2003) Roles of systemic intermediaries in transition processes. Int J Innov Manag 7:247–279

Chapter 2
Place in Transitions—Concepts for When it Matters: Essentially, Accidentally, Locus and Nexus

Fjalar de Haan

Abstract How much does place matter in transitions? And if it matters much, how to deal with it? This research essay explores the conceptual aspects of these issues. At the core is that transitions concepts and frameworks typically employ a functional systems perspective in which place and scale are implicit. The burgeoning literature on the geography of transitions and urban transitions provides many clues as to the aspects of place and scale that would be of conceptual import for transitions. When can a transition truly be considered *of*—rather than 'merely' occurring in, or to—a certain place? I argue that transitions can be considered to cover a spectrum ranging from *accidentally* place based—when place matters only because things need to happen *somewhere*—and *essentially* place based—when the transition dynamics are completely contingent on the local context, needs and aspirations. For transitions erring on the essential end of the spectrum, I argue, a systemic conceptualisation of place—which I call the *locus*—is useful, while the systems affected by the transitions, the *nexus*, can be conceptualised in the familiar ways. Locus and nexus are mutually embedded, and an analytical interface is found in the networks and interactions of actors.

Keywords Geography of transitions • Urban transitions • Place • Locus • Nexus • Essentially place based • Accidentally place based

2.1 Introduction

The role of place and scale in transitions has become an important aspect of sustainability transitions research. While before, arguably, matters of place and scale were under appreciated in the field, geographers of transitions and urban transitionists have put these matters firmly on the research agenda. This will no

---

F. de Haan (✉)
Melbourne Sustainable Society Institute and Melbourne School of Design, The University of Melbourne, Melbourne, VIC, Australia
e-mail: fjalar.dehaan@unimelb.edu.au

T. Moore et al. (eds.), *Urban Sustainability Transitions*, Theory and Practice of Urban Sustainability Transitions, DOI 10.1007/978-981-10-4792-3_2

doubt draw increased attention to spatial concerns, as in fact it has already done. But sheer research attention as such may not necessarily improve the appreciation of the role of place and scale in the *understanding* of transitions. As a theoretician, I think this is because the issue at hand is to an important degree a conceptual one. The conceptual portfolio of transitions study is, at this point, not very well equipped to deal with spatial matters, approaching the subject matter from a functional point of view instead.

Therefore this research essay starts with an investigation into the current conceptual challenges regarding matters of place and scale in transitions theoretical concepts. I will then explore the possibilities of accommodating a spatial conceptualisation amongst the present conceptual portfolio—that is, without discarding or tinkering with the available useful conceptual tools—and propose a framework to integrate them, based on ideas of how the spatial and the functional come together in change agents.

As part of this, I will first explore why there would be any conceptual challenges to begin with. Most applications of the canonical frameworks, likely even at the present time, only have reference to place in the spatial delineation of the case study—it has to happen *somewhere* rather than elsewhere after all.[1] In many cases, this seems a satisfactory mode of analysis, at least to a degree. Yet it is also clear that in some cases, place is a crucial determinant in the dynamics of a transition—when the directions and aspirations for change are intimately connected with the local culture and physical geography, for instance. Such cases, which I will refer to as *essentially* place based, need adequate conceptualisation of their spatial aspects, while transitions that are only *accidentally* place based may be well understood within a more spatially agnostic framework.

The need for spatial conceptualisation notwithstanding, there are sound reasons to continue to embrace a functional perspective on transitions. I will discuss these later in the essay, but in the meanwhile it is evident to most transitions researchers that the systems affected by transitions, systems of resource and service provision, are not at all necessarily located at, or confined to, the places where their benefits are reaped or damages are suffered. Health care, though locally practised, typically depends on national legal and financial arrangements. Similarly, the food consumed in cities is mostly not produced within their topographical boundaries. In other words, even for essentially place-based transitions, it is necessary to regard the functional system or systems affected. This aspect, the bundle of implicated systems, I will refer to as the *nexus*. Conversely, even for a purely accidentally place-based transition, a conceptualisation of that place is required. I will refer to that aspect as the *locus*. With these concepts I aim to address the need for 'a conceptual vocabulary to be added to transition analyses in order to better contextualise them in "real spaces"', formulated by Coenen et al. (2012, p. 973).

Working from this dualist—locus and nexus—perspective on transitions, I will elaborate (1) a proposal for a conceptualisation of the locus and (2) hypotheses

---

[1] Though they could also, in a certain sense, happen *everywhere*.

on the connection and interactions between locus and nexus. The former entails a systemic conception of place, with a non-arbitrary spatial delineation as recognised by actors that form part of that system. Put plainly, an area people give a name to because it can meaningfully be referred to as a unit. Cities are of course prime examples—and the focus of this book. Such a primary delineation then forms the core of understanding that system as embedded in levels of governance and various social, economical and political networks. These, in turn, allow the hypotheses connecting the locus and the nexus to be framed in terms of actors, mediating the place-specific needs and aspirations and the systems they rely upon.

2.2 Transitions Theories Have No Place

Transitions theories study transformations in systems of provision of goods and services, at various levels of detail. The Multi-Level Perspective (MLP, see, e.g. Rip and Kemp 1998; Geels 2002; Geels and Schot 2007), for example, looks at the dominant technology in performing some societal function and the socio-technical configuration around it. The Multi-Pattern Approach (MPA, see, e.g. de Haan 2010; de Haan and Rotmans 2011) looks at the various subsystems, referred to as constellations, that may be involved in meeting certain societal needs. Technological Innovation Systems (TIS, see, e.g. Jacobsson and Johnson 2000; Hekkert et al. 2007; Suurs 2009), in slight contrasts, looks at the systems in which new technologies[2] emerge that, in turn, may lead to changes in the systems such as described by the MLP and MPA.

These frameworks all have a *functional* view on their objects of study. Functional can be understood in two senses: (1) looking at the function a certain system may have in society, i.e. functionalist, or (2) looking at the functioning of these systems, e.g. to understand why they need this or that element to function (in the previous sense) or how they can be changed.[3] Both senses are present in the referred to frameworks. A functional perspective is very appropriate in the context of transitions studies. An important reason to study transitions is to overcome persistent societal problems, like sustainability-related issues, and such problems are precisely thought to be intimately related to the way the relevant systems function (see, e.g. Schuitmaker 2012; Rotmans 2005). But in a functional analysis, it may not matter much *where* a certain element contributes its part to the overall functioning. Moreover, where a functional system begins or ends in a *spatial* sense is

[2]It does not really need to revolve around technologies; one could also talk about 'practices', for example. But many, probably even a large majority of, cases discussed with particularly the MLP or TIS are about technologies.

[3]See, e.g. Cummins (1975) or Mahner and Bunge (2001), for a nuanced understanding of different kinds of functional explanation.

not of primary importance. No surprise then that the concepts of transitions studies are spatially agnostic.

In the case of TIS, this agnosticism is a conscious choice. The perspective evolved out of an approach that did have an explicit geographical delineation, namely, National Systems of Innovation,[4] and the reason for this evolution would have been the realisation that systems of innovation often do not respect national boundaries. The development of the MLP shows similar considerations. In earlier work, especially by Geels (e.g. 2002), the cases are clearly delineated in space and time.[5] The concepts may be agnostic, but the change happens somewhere. In work of Geels with Raven (2006) that already tries to accommodate a sense of space, however, the MLP concept of 'niche' is split in several local, i.e. spatially based, manifestations contributing to an a-spatial, 'global' counterpart. These examples serve to show that a functional, a-spatial view, is not necessarily a shortcoming and has, on the contrary, been introduced for theoretically sound reasons.

But, as mentioned, it comes with some conceptual issues. Take, for example, the very idea of an 'urban transition'. If a transition is a fundamental shift in one of those functional systems referred to earlier—energy supply, mobility, health care, etc.–then many transitions occurring to cities cannot be called 'urban' in a very meaningful way. As Horne explained in Chap. 1 of this volume, some transitions happen to cities without being urban transitions in any distinctive sense, that is, they are only urban to the extent the affected systems are located within the cities' boundaries. Health care transitions, then, are rarely urban (as alluded to before, the relevant institutional aspects would typically be national), while transitions in water management such as drinking water supply or sanitation, often are, as their infrastructures as well as the people they service reside in the urban area. In both of these extremes, the urban character of the transition is in a way coincidental. This is what I referred to earlier as a transition being *accidentally* place based, in this case accidentally urban.

Are there, then, by contrast, transitions that are *essentially* place based? When can a transition truly be considered *of*—rather than 'merely' occurring in, or to—a certain place? Transitions that are the consequences of the particular needs and aspirations of a particular place? Unfolding in a way contingent on the local context and history? In the next section, I will distil some lessons about the importance of place as suggested by geographers of transitions. But as a prelude, I observe that cities at least seem to aspire to transitions. Hodson and Marvin (2010, pp. 480–481) point this out, as do Loorbach and Shiroyama (2016, pp. 5–6). Aspiration suggests agency, and Loorbach and Shiroyama (loc. cit.) refer to the insight that cities can be 'actors and locations' of sustainability transitions. Perhaps it is a hallmark of an essentially place-based transition that agency and scenery be co-located. Moreover, cities' transitional aspirations are often formulated in rather

[4]See also the discussion by Coenen et al. (2012) of the development of TIS out of the National Systems of Innovation approach.

[5]This is still the way most MLP analyses are conducted.

broad terms, like 'sustainability', 'climate neutrality', 'resilience' or 'liveability'. Terms that suggest that indeed a whole bundle of functional systems related to the city needs to undergo transitions.

Starting in this fashion, from the needs and aspirations of a specific place, then seems to yield the complementary issue that there is no longer a specific functional system to serve as transition protagonist. An analytical way out of this would be to introduce a more ad hoc, problem-focussed systems analysis which would incorporate the relevant aspects of potentially several systems—as well as the relevant particulars of the place. This latter option is indeed what DRIFT researchers Roorda et al. (2014) suggest in their guidance manual for Transition Management in urban contexts. Transition Management (TM, see, e.g. Loorbach 2010) is a full-fledged transitions-theoretical perspective in its own right, but unlike the MLP or TIS, it is concerned with the organisation of a transition as a process, driven by actors. Systems analysis is a crucial part of TM as an approach, but although TM builds upon and extends insights from other transitions theories, it can be flexible about what framework to use for its systems analyses. Thus, in the aforementioned—tried-and-tested—guidance manual, they suggest a stocks-and-flows integrated sustainability systems analysis, that is, *not* one of the transitions systems frameworks.

It is of course also possible to use a functionally oriented framework, such as the MLP, as a starting point for a spatially informed analysis. This is what Raven et al. (2012) suggest. Their approach—in simplified, practical terms—is to start from some spatio-functional case delineation, identify the relevant regimes and niches and then investigate how the spatial relations influence the transition dynamics. They tentatively apply this to a case study of biomass gasification in India which shows regional differences in uptake of this technology. The suggestion is that spatially informed explanations of this would entail investigating spatial relations through the networks of the relevant actors and territorial disaggregation of the MLP concepts (e.g. looking at the state-level energy regimes).

Surely this is a sophistication over the more usual analyses where space remains implicit. This kind of approach does however seem to assume that any spatial specificity is *accidental* in my sense of the word. In other words, the transition is there, or rather, just *is*, and its dynamics are in various degrees locally specific or determined. This reflects the idea—imported from relational geography—that 'no a priori privilege is given to any scale' (Hansen and Coenen 2015 referring to Boggs and Rantisi 2003). For an *essentially* place-based transition this idea may actually be misgiven as for such transitions a 'privileged', or at least defining (rather than defined), scale should be evident—for example in shaping the needs and aspirations for change. This—to me—indicates that space, place and scale as such need to be conceptualised in a systemic way, beyond just providing another 'dimension' alongside time and structuration.

In the coming sections, I will work towards such a systemic conceptualisation of place in transitions and its connection to the implicated functional systems. Not being an expert on matters of place, I will prepare by trying to learn from those who are.

2.3 Geography Lessons

The literature on the geography of transitions seems very convinced, not only that place matters but that it matters *a lot*. Hansen and Coenen (2015, p. 95), for example, state that ‘Transitions are constituted spatially’ and ‘Sustainability transitions are geographical processes’. Such statements suggest deep metaphysical truths about transitions, but I am personally more inclined to think they reveal the natural inclination of geographers to view any phenomenon as essentially place based. Their point, however, is well taken, and my rendering of it is this: transitions are—even if only in the last instance—the consequences of human actions, and human actions are shaped by the context in which those humans operate, which is, in turn, at least to a degree geographically determined. So, how much does place matter for transitions?

Surely, there are cases where place and scale are crucially important for the dynamics of transitions. Regional reorientations would be clear examples, such as the early Transition Management case (see, e.g. Loorbach and Rotmans 2010) in the South-East of the Netherlands, a former mining region struggling to find a new identity. Urban transition processes such as described by Roorda et al. (2014) would also be very contingent on the local context. Surely also, there are transitions cases where the local context is all but irrelevant. This is likely anathema to the geographers who may be reading this, but bear with me. Local context may matter very little in two—not particularly rare—circumstances: (1) there simply is no relevant spatial scale and change is homogeneously distributed through e.g. legislative reform, and (2) a new technology or practice emerges ‘globally’ and finds uptake in various places. These are more likely to change the local context than the other way around.

So perhaps the question should instead be in what cases place matters much in transitions, as there seems to be a spectrum ‘measuring’ the importance of the local context. At one extreme end of this spectrum, one finds transitions that are only *accidentally* place based (because they have to happen *somewhere*) while at the other extreme one finds transitions that are *essentially* place based (because there, place determines the why and how of the transition). Whether any empirical case actually occupies one of these extremes is perhaps besides the point, what matters is to identify what matters about place in the cases where it does. Once that is clearer, I have a starting point for a conceptualisation of the *locus* of transitions. If anyone, it is the geographers and urban transitionists who know what matters about place in transitions, so I will gratefully appropriate their findings.

As a basis for the conceptualisation of the locus and the connection between locus and nexus, I will particularly draw lessons from two sources, Hansen and Coenen’s (2015) article which—amongst other things—surveys the literature on place and transitions and provides a very useful bit of meta-analysis and Hodson and Marvin’s (2010) article from which a suggestion regarding the locus-nexus coupling can be distilled. I will start with the former.

In their survey of the literature on the influence of space on transitions, Hansen and Coenen (2015) identify six recurring themes, which I will use as a proxy. The themes are 'urban and regional visions and policies', 'informal localised institutions', 'local natural resource endowments', 'local technological and industrial specialisation' and 'consumers and local market formation'. Under these headers, Hansen and Coenen discuss various insights and emphases of a representative, if not exhaustive, range of authors. I will take the liberty of reinterpreting and reorganising these themes to serve as the basis of a systemic conceptualisation of place. I suggest the following three broad 'dimensions' along which place may matter to transition dynamics, in other words, dimensions of the locus:

- Governance and Institutional Arrangements
- Cultural Setting and Social Practices
- Physical Geography and Natural Resources

Governance and Institutional Arrangements roughly coincide with the things discussed under the theme of 'urban and regional visions and policies' by Hansen and Coenen. By 'governance' I mean the usual, i.e. the official governmental structures pertaining to the place as well as the broader arrangements to do with organising public and economic matters. Institutional arrangements are particularly meant to refer to formal institutions like laws, regulations and policies but also include established alliances and collaborations and such. Informal institutions are considered under *Cultural Setting and Social Practice*. This term I introduce to explicitly refer to local cultural aspects including language and practice aspects. These aspects, in slightly different terms, are covered by Hansen and Coenen under the theme 'informal local institutions'.

Lastly, I introduce the category *Physical Geography and Natural Resources*. Interestingly, Hansen and Coenen's survey suggests physical geography is not a theme in the literature they include but for the 'local natural resource endowments'. I would argue, however, that in many transition cases, the local physical geography is a very relevant factor. Think, for instance, about the different consequences places can suffer from climate change and how that would matter for adaptation measures. Think also how climate influences the potential for certain technological solutions, such as solar photovoltaics or rain- and stormwater harvesting. A more classic example is how water infrastructure (supply, drainage and sewerage) has to take into account the local physical geography. Healthcare provision in remote areas requires different servicing facilities and so on.

It would seem as if I just discard the last two themes identified by Hansen and Coenen, and this is partially true. Of 'consumers and local market formation', I consider the local market aspects to be part of the *Governance and Institutional Arrangements*, while I consider consumers to be part of both this category and *Cultural Setting and Social Practices*. Similarly, aspects of 'Local technological and industrial specialisation', I consider under *Governance and Institutional Arrangements* and others under *Cultural Setting and Social Practices*. This is not to downplay the importance of these last two themes in the localised emergence of environmental and social innovations and in turn their role in transitions, but the conceptualisation I am after is about place and transitions, not place and innovation. It is a common predilection of transitions studies—and the geographical approaches seem no exception—to treat transitions as matters of *innovation* rather than *transformation*.[6] Innovation can certainly be an important part of transition dynamics, but here I think that giving these aspects separate categories unduly shifts the emphasis.

In the coming section (Sect. 2.4), I will use these three dimensions to further conceptualise the locus of a transition and elaborate somewhat on how they enable treatment of scale and embedding in different ways. In the section following that (Sect. 2.5), I will elaborate on the connection between the locus and the nexus and argue that this connection is to be found in actors. Incidentally, actors will also be central in establishing a meaningful conception of a place as a system. This has to do with the notion that actors define the relevant scales, with scale understood more broadly than physical distance only, for example, as degrees of separation in social networks or other notions of proximity and distance such as those in power relations—a lesson learnt from Coenen et al. (2012, p. 969).

Having actors in the conceptual portfolio will provide a more straightforward way to think about *purposive* transitions. In that context, special attention to those actors that (potentially) connect the locus and nexus seems appropriate. A cue to that effect can be taken from Hodson and Marvin (2010, p. 482) whose loci are cities, when they say that '[v]isions of purposive urban transitions represent a transformative view of the relationship between cities and sociotechnical regimes. [...] With purposive urban socio-technical transitions, therefore, the aim is to mutually transform both urban governance regimes and socio-technical regimes', particularly, when realising they also talk about urban governance *networks* there—the governance network of the locus, as it were. To me the networks related to the nexus are equally important. Where these networks (should) interconnect, I argue, there will be opportunities for purposive transitions, and this will allow me to connect to concepts and proposals from Transition Management.

2.4 Locus and Nexus: Place and What It Brings Together

Locus and nexus are two not mutually exclusive aspects of a place. Locus refers to all that is particular to a place as a spatial configuration including its socio-historical evolution. Nexus refers to the bundle of systems that provides the resources and services the place depends on for its functioning. A biological, or even medical, metaphor may serve to illustrate the relation and distinction between the two

[6]Hansen and Coenen (2015, p. 104) do acknowledge and discuss this, observing that 'the greater majority of the studies have focused primarily on the geography of niche developments and formative phases in technological innovation systems' and that the broader literature has 'been infatuated with a "bottom-up" approach to transitions that have primarily considered niche-based processes'.

concepts. The locus is the body as an entity contiguous in space and time to which its history and identity attach. The nexus is the body as the functional collection of organs, producing the metabolism it needs to be alive. The locus is the substrate of the nexus, the nexus sustains the locus.

The more the dynamics of a transition is contingent upon elements of these dimensions—that is, the more it is determined by the locus—the more it is an essentially place-based transition. Conversely, the more the transition dynamics is determined by the composition of the functional systems implicated in the transition—that is, the more it is determined by the nexus—the more it is accidental to the where the transition takes place.

That, roughly, is the framework of the relation between transitions and place I am trying to convey. A framework in which place itself is considered a systemic entity—the locus—that relies on other systems to support itself. Those other systems—the nexus—provide resources or services and are therefore the functional systems we comfortably conceptualise within the usual frameworks from transitions studies.

Using the term 'nexus' the way I do here, to denote a bundle of functional systems, is of course not completely new. Also in human geography, the term nexus has attracted interest, in particular referring to the bundle of systems related to water, energy and food provision (e.g. Rees 2013; Leck et al. 2015) it seems, though the term is also used to refer to other bundles such as resource-state (Bridge 2014) which may correspond less to the stricter interpretation I give the term.

Before I elaborate the locus-nexus framework, I need to acknowledge and deal with some conceptual issues it introduces or reintroduces in altered form. To do this it is illustrative to see how the corporeal metaphor breaks down. It does so fairly directly in two ways:

1. A body is the paradigm case of a system with functional parts, in fact, much of functionalist social theorising seems to have fruitfully drawn from this metaphor. A place, however, is not obviously a system at all, At least not just any place. If I would have substituted 'city' for 'place' in the metaphor, attributing systemhood to it would not readily be perceived as absurd. Similarly perhaps for other 'naturally' delineated spatial territories such as islands or ecosystems. But not every delineation seems to automatically define a system. In other words, the issue at hand is to establish criteria for a place to meaningfully be considered a system.
2. Unlike the organs in a body, the systems that sustain a place are not necessarily confined to its boundaries. Cities' vital systems, for instance, perhaps even more often than not, extend beyond the spatial territory of the cities proper. Electricity grids are a point in case. I mentioned this in the introduction already, with examples such as agriculture, which feeds cities but takes place outside of them, and health care which usually relies on national institutional structures. So far, this is not a problem, and it is moreover, one of the issues I think the locus-nexus view clarifies to an extent. However, systems in the nexus *may* be manifestly local, but they need not be.

To start with the latter, this seems to be the original problem all over again. Fortunately, the question of where the various parts of a system reside is less complex than the question of where the various influences and implications of a transition are. In fact, a spatial analysis of a functionally conceptualised system is a ready possibility and I will suggest and elaborate something along those lines when connecting locus and nexus in the next section. Regarding the former, a meaningful conception of the locus as a system is precisely what I mean to elaborate next.

2.4.1 The Locus

Not every place is a system, I seemed to have claimed. This is however, in some sense a matter of philosophical, even metaphysical, disposition. It is, to my mind, perfectly reasonable to analyse some empirical phenomenon, or even conceptual structure, as a system, without caring one bit about its ontological status—whether it exists, as a system, out there. Much in the same way the physicist does not have to 'believe' in the electric field to appreciate its explanatory power.[7] In such an analysis, a system need be nothing more than a set of elements united by a set of relations which may have been imposed by the analyst.

In some cases, however, the elements of a system seem to be aware of these relations, and the system may be said to be aware of its systemhood, as it were. This may come across as an almost mystical statement, but it is far from that. Simple and straightforward examples are social groups such as nuclear families, companies, etc. More complex examples would be economies or academic disciplines. In all these examples, the elements—here, individuals—are aware of the systems they form part of. Moreover, often the rules and dynamics of the system are produced by the elements, which at the same time abide by those rules and dynamics. Such a connection between 'lower'- and 'higher'-level phenomena is commonly referred to as 'emergence' and for the cases where it depends on agents with reflective capacities such as some humans, I have called it 'reflective' or 'type-III' emergence (de Haan 2006).[8] In sociology this is sometimes referred to as the duality of structure and agency, after Giddens (1979, 1984).

It is in this sense that I propose to consider a locus a system, and in that sense not every place is a system. Many interesting categories of place, however, are: cities, ports, regions, states, federations, neighbourhoods. Even smaller delineations such

[7]This analogy is less contrived than it may appear. The electric field and other classical fields in physics, were first introduced as mathematical fictions to facilitate otherwise cumbersome calculations. The physically 'real'—out there—quantity on which the mathematical fiction was based is electric *charge* in this example.

[8]Reflective agents are by no means a prerequisite for emergence. There are myriad examples of emergence without, for example in physics, chemistry, biology and computer science. The kind of criterion I have for systemhood, in the sense here intended, would be something like what Crutchfield (1994a,b) referred to as 'closure of newness'.

as squares or buildings, in certain circumstances, can meaningfully be considered systems. This is obvious from the fact that in many such places, special rules apply, such places have names and, to push the point, the corporeal metaphor seems to suggest itself, to the point that personifications are frequently observed.

That loci delineate themselves is of course not unknown to geographers nor to the geographers of transitions, it is part of the phenomenon that 'scales are actively constructed [. . .] by actors' (Coenen et al. 2012, p. 976). In Sect. 2.3, I distilled three dimensions along which place may matter in transitions, so it does not seem far-fetched to assume these are the dimensions along which loci would be delineated. Or, as some would prefer to phrase it, those may well be the dimensions along which actors 'construct scale'. For the first dimension (Governance and Institutional Arrangements), this seems fairly obvious, especially for urban areas such as the City of Melbourne or the Port of Rotterdam, but also e.g. the eurozone. The second (Cultural Setting and Social Practices) perhaps not so for cities, but all the more for certain regions, like the Bible Belt or Silicon Valley. The latter can of course also be interpreted as a delineation along the third dimension (Physical Geography and Natural Resources) as its name refers to an actual valley in California, USA (Santa Clara Valley) though the spatial area associated with the name Silicon Valley now also includes other areas.

In summary, systemic places—loci—emerge (in the complex systems meaning of the word) as the intended or inadvertent result of the doings of actors. The three dimensions can be used—heuristically—to probe the contours of such places. In a similar fashion to how actors produce systemic places, they embed them, connect them and subdivide them. And the three dimensions can therefore also be used in a similar fashion to map the various spatial relations that matter. This way of looking at actors and how their relations and interactions produce the relevant scales is of course akin to the orientation in geography dubbed the 'relational turn' (Boggs and Rantisi 2003). Scholars taking that turn (e.g. Amin 2002) also emphasise that the spatial dimension need not always be the key dimension[9] in geographic analyses and that, for example, social relations can be the basis of notions like connection, proximity and distance. By viewing actors as part of networks affiliated with spatial systems—loci—as well as with the networks of functional systems—the nexus—a useful conceptual connection between locus and nexus can be established. Actors establish the mutual embedding of locus and nexus.

2.5 Connecting Locus and Nexus: Mutual Embedding and Actors

The first step to connecting locus and nexus is to appreciate their mutual embeddedness. Although the separation seems clean and clear, there are aspects of place to be found in the nexus' systems and functional dependencies in loci—the nexus is embedded in the locus and the locus is embedded in the nexus.

[9]Apologies for the muddled multiple usage of the term 'dimension' here.

Why does a locus depend on functional systems at all? Because it has needs. The people living, working or otherwise interacting with it have needs, the structures, be they ecological, institutional or technical, have needs and so forth. The latter may be derivative needs of the former, but the point remains there are needs to be met in any locus. Such needs are societal needs and they are of course the very reason functional systems like those of the nexus exist to begin with (de Haan et al. 2014). Moreover, much of the Governance and Institutional Arrangements dimension of a locus is devoted to making sure societal needs are met and will continue to be met. Changes—perceived, anticipated or acute—in societal needs and circumstances give rise to *aspirations* that are another important aspect of locus governance.

What is connected to space in a nexus' systems? The functional systems I discussed are systems of resource or service provision. In as much as these systems provide—meet the needs of—the locus, there must be some connection between them and it.[10] Also, the functioning of the systems in a nexus depends on social, technical and biophysical infrastructures, any of which may reside within the perimeter of the locus—it may even be the case that such infrastructures reside within a different locus than the one or ones they service.

As discussed, systems in a nexus usually have their own accompanying governance and institutional arrangements. Then—seeing that the governance of a locus is to an extent governance of its relation with its nexus—an interesting analytic question is to what extent their Governance and Institutional Arrangements dimensions overlap. Conversely—seeing that some of the social, technical or biophysical infrastructures of the nexus may reside within the locus—the question is to what extent those aspects are shared amongst locus and nexus. In other words, for each system in the nexus of any particular locus, one can analyse the degree of institutional (the former) and territorial (the latter) control and responsibility the locus has over the nexus *and vice versa*. It is the analytic question of the power balance between a place and the systems it depends on.

The second step in connecting locus and nexus—both conceptually and practically—is by identifying the relevant actors and their roles in both. In as much as the locus and nexus are mutually embedded, that is, where they systemically intersect, finding the connecting actors is relatively straightforward. Identify the actors that have a formal *role* in the intersection of locus and nexus. Not all relevant actors can be found in this way of course, and of particular interest are those actors who move between systems—as part of their roles or otherwise—and those who challenge the very structures such roles are part of.

The former, referred to as *boundary spanners*, are actors who are strongly related to the institutional aspects—the rules and organisation enabling the functioning—of

[10]The resources, services and labour going into these functional systems can of course also constitute a connection with the locus, just as the waste flows and other adverse side effect they may produce. These aspects are of course very important from a sustainability point of view, but I will take the liberty of only skimming over them in this fashion.

particular systems, but who may also have similar ties with other systems or are able to 'work' between or across them. A particularly useful conceptualisation[11] of such boundary-spanning actors is given by Brodnik and Brown (2014). They build upon an institutional logic perspective on transitions developed by Fuenfschilling and Truffer (2014) and conceptualise functional systems as 'constellations of logics'—where the term 'constellation' is no doubt intentionally suggestive of my constellation conception of societal systems (see, e.g. de Haan and Rotmans 2011) as it refers to the same kind of system. Boundary-spanning actors in this picture are individuals able to 'bridge, blend or transpose' aspects of the institutional logic (Brodnik and Brown 2014) which can contribute to the transformation of the systems based on these logics.

Some actors are not so much defined by the institutions they represent or enact—in work with Jan Rotmans (e.g. de Haan and Rotmans 2015, 2016), I refer to such actors as 'institutional actors'—but by the institutions and systems they aim to change or the alternatives they advocate. These latter actors we referred to as *transformative actors*, for obvious reasons. The distinction between institutional and transformative actors is a conceptual one, and an individual may at the same time be an institutional actor and a transformative actor. In our typology of transformative actors, we therefore also refer to those types as 'roles'. Indeed, it may greatly benefit the transformative agenda of an actor if she also is an institutional actor, in which capacity she may have access to useful transformative levers. And, indeed also, a transformative actor may be a boundary spanner and more successful for it.

2.6 Conclusions: Purposive Transitions and When Place Matters

I would like to conclude this research essay by suggesting two areas where the key concepts presented may prove useful. The first is in the ongoing conceptual development of the geography of transitions and the second is Transition Management and other approaches to pursue purposive transitions. I will begin with the former.

Truffer et al. (2015) name socio-spatial embedding as one of the three main conceptual dimensions of a research agenda for the geography of transitions, along with multi-scalarity and issues of power. For them, socio-spatial embedding refers to the conditions that make certain places more transition prone than others. Of course, in this something shines through of an implicit assumption that the prime determinants are in fact of a geographical nature (see also Sect. 2.3). The spectrum from essentially to accidentally place-based I proposed provides a way to probe the 'depth' of a transition's socio-spatial embedding, turning the implicit assumption into a potential research question. When the transition of interest errs on the essential

[11] See also Smink et al. (2015) for a recent treatment of boundary spanners in the context of transitions.

end of the spectrum, the concepts of locus and nexus can be employed to separate the analysis of place and implicated functional systems so as to more explicitly appreciate the place-based factors in the transformative dynamics.

The depth of a transition's socio-spatial embedding has an additional bearing in the context of purposive transitions. A transition may, for example, require a large degree of embedding in particular places to take off—in this essay's terminology, the transition needs to be essentially of those places. Conversely, successful transition cases may remain local exceptions because they are too deeply embedded there—the transition is too essentially place based. The importance of governing the transition on this spectrum is especially pertinent when cities are concerned.

The sustainability transitions scholarship has been giving more and more attention to cities in recent years. Cities are seen as incubators or generators of the social and technical innovations required for sustainability transitions. Loorbach and Shiroyama (2016, p. 8) speak of cities as 'transition machines' in this way and also see a role for them in accelerating transitions at larger scales. They, as do Hodson and Marvin (2010), observe that cities also want to be protagonists in transitions and mention several initiatives and organisations embodying such ambitions. Hodson and Marvin, p. 480 also note that cities may 'receive' a transition from the national level, which they can possibly 'mediate'. No wonder perhaps then that Wittmayer and Loorbach (2016) signal an 'urban turn' in Transition Management'. Indeed, the book and book series this essay is part of are testament to the interest of the sustainability transitions field in 'the urban'.

This 'urban turn' can be understood as an *analytic* shift to understand urban transitions as essentially place-based. In that sense, the turn is similar to the 'geographic turn' in transitions studies more broadly, so, mutatis mutandis, the spectrum essential-accidental can be put to use. Also, the locus-nexus distinction seems particularly suitable to cities. Cities readily allow themselves to be seen as loci and almost invite a systemic view on them—it is not coincidental they and 'the urban' featured as convenient examples throughout this essay. Moreover, it is in cities that one readily appreciates the mutual embeddedness of the locus and the nexus, as urban life is sustained by many often interrelated functional systems.

As Transition Management is interested in forming alliances of transformative actors, it may find use for the locus-nexus distinction as a framework. For example in connecting actors representing local needs and aspirations to representatives of certain functional systems, whose geographical perspectives may entail different scales and other loci. The essential-accidental spectrum provides a perspective to investigate the functioning of a city as a Loorbach and Shiroyama 'transition machine' as it were, or to explore its potential to become one. Similarly, in such cases as Hodson and Marvin (2010) noted where cities 'receive' transitions from national or supra-national levels, a—potentially important—part of the governance of the transition would be to make it essentially place based.

Acknowledgements I would like to thank Lars Coenen, Andréanne Doyon, Ralph Horne and Trivess Moore for the various ways in which they helped me improve this chapter and the concepts presented in it.

References

Amin A (2002) Spatialities of globalisation. Environ Plan A 34(3):385–399

Boggs JS, Rantisi NM (2003) The 'relational turn' in economic geography. J Econ Geogr 3(2):109–116

Bridge G (2014) Resource geographies II: the resource-state nexus. Prog Hum Geogr 38(1):118–130

Brodnik C, Brown RR (2014) Transformative change agency in sustainability transitions: the interplay of boundary spanning and institutional logics in Melbourne, Australia, Utrecht University, The Netherlands

Coenen L, Benneworth P, Truffer B (2012) Toward a spatial perspective on sustainability transitions. Res Policy 41(6):968–979

Crutchfield JP (1994a) The calculi of emergence: computation, dynamics and induction. Phys D: Nonlinear Phenom 75(1–3):11–54.

Crutchfield JP (1994b) Is anything ever new? Considering emergence. In: Cowan G, Pines D, Melzner D (eds) Integrative themes. Number XIX in Santa Fe Institute Studies in the Sciences of Complexity. Addison-Wesley, Reading/Santa Fe

Cummins R (1975) Functional analysis. J Philos 72(20):741–765

de Haan FJ (2006) How emergence arises. Ecol Complex 3(4):293–301

de Haan FJ (2010) Towards transition theory, Ph.D. thesis. DRIFT (Dutch Research Institute for Transitions) and Faculty of Social Sciences, Erasmus University Rotterdam, Rotterdam

de Haan FJ, Ferguson BC, Adamowicz RC, Johnstone P, Brown RR, Wong THF (2014) The needs of society: a new understanding of transitions, sustainability and liveability. Technol Forecast Soc Chang 85:121–132

de Haan FJ, Rotmans J (2011) Patterns in transitions: understanding complex chains of change. Technol Forecast Soc Chang 78(1):90–102

de Haan FJ, Rotmans J (2015) Beyond Current transitions theory – facing the next great transition, University of Sussex, Falmer, Brighton

de Haan FJ, Rotmans J (2016) They make the change – roles of actors in transitions, MSSI research paper no. 5, Melbourne Sustainable Society Institute, The University of Melbourne. ISBN:

Fuenfschilling L, Truffer B (2014) The structuration of socio-technical regimes—Conceptual foundations from institutional theory. Res Policy 43(4):772–791

Geels FW (2002) Technological transitions as evolutionary reconfiguration processes: a multi-level perspective and a case-study. Res Policy 31(8–9):1257–1274

Geels F, Raven R (2006) Non-linearity and Expectations in Niche-Development Trajectories: ups and downs in Dutch Biogas Development (1973–2003). Technol Anal Strateg Manag 18(3–4):375–392

Geels FW, Schot J (2007) Typology of sociotechnical transition pathways. Res Policy 36:399–417

Giddens A (1979) Central problems in social theory: action, structure and contradiction in social analysis. University of California Press, Berkeley

Giddens A (1984) The constitution of society. University of California Press, Berkeley/Los Angeles

Hansen T, Coenen L (2015) The geography of sustainability transitions: review, synthesis and reflections on an emergent research field. Environ Innov Soc Transit 17:92–109

Hekkert MP, Suurs RAA, Negro SO, Kuhlmann S, Smits REHM (2007) Functions of innovation systems: a new approach for analysing technological change. Technol Forecast Soc Chang 74:413–423

Hodson M, Marvin S (2010) Can cities shape socio-technical transitions and how would we know if they were? Res Policy – Spec Sect Innov Sustain Transit 39(4):477–485
Horne R (2017) Urban low carbon transitions: housing and urban change, in 'Chapter 2. Urban Sustainability Transitions: Australian Cases – International Perspectives'.
Jacobsson S, Johnson A (2000) The diffusion of renewable energy technology: an analytical framework and key issues for research. Energy Policy 28:625–640
Leck H, Conway D, Bradshaw M, Rees J (2015) Tracing the water-energy-food nexus: description, theory and practice. Geogr Compass 9(8):445–460
Loorbach D (2010) Transition management for sustainable development: a prescriptive, complexity-based governance framework. Governance 23(1):161–183
Loorbach D, Rotmans J (2010) The practice of transition management: examples and lessons from four distinct cases. Futures 42(3):237–246
Loorbach D, Shiroyama H (2016) The challenge of sustainable urban development and transforming cities. In: Loorbach D, Wittmayer MJ, Shiroyama H, Fujino J, Mizuguchi S (eds.) Governance of urban sustainability transitions: European and Asian experiences. Theory and practice of urban sustainability transitions. Springer Japan, Tokyo, pp 3–12
Mahner M, Bunge M (2001) Function and functionalism: a synthetic perspective. Philos Sci 68(1):75–94
Raven R, Schot J, Berkhout F (2012) Space and scale in socio-technical transitions. Environ Innov Soc Transit 4:63–78
Rees J (2013) Geography and the nexus: presidential address and record of the Royal Geographical Society (with IBG) AGM 2013. Geogr J 179(3):279–282
Rip A, Kemp R (1998) Technological change. In: Rayner S, Malone EL (eds) Human choice and climate change, vol 2. Battelle Press, Columbus, pp 327–399
Roorda C, Wittmayer J, Henneman P, van Steenbergen F, Frantzeskaki N, Loorbach D (2014) Transition management in the urban context: guidance manual, Technical report, Dutch Research Institute For Transitions (DRIFT), Erasmus University Rotterdam, Rotterdam
Rotmans J (2005) Societal innovation: between dream and reality lies complexity. Inaugural addresses research in management Series. ERIM, Erasmus Research Institute of Management, Rotterdam. Inaugural address
Schuitmaker TJ (2012) Identifying and unravelling persistent problems. Technol Forecast Soc Chang 79(6):1021–1031
Smink M, Negro SO, Niesten E, Hekkert MP (2015) How mismatching institutional logics hinder niche–regime interaction and how boundary spanners intervene. Technol Forecast Soc Chang 100:225–237
Suurs RAA (2009) Motors of sustainable innovation: towards a theory on the dynamics of technological innovation systems, PhD Thesis. Utrecht University, Utrecht
Truffer B, Murphy JT, Raven R (2015) The geography of sustainability transitions: contours of an emerging theme. Environ Innov Soc Transit 17:63–72
Wittmayer JM, Loorbach D (2016) Governing Transitions in cities: fostering alternative ideas, practices, and social relations through transition management. In: Loorbach D, Wittmayer MJ, Shiroyama H, Fujino J, Mizuguchi S (eds) Governance of urban sustainability transitions: European and Asian experiences. Theory and practice of urban sustainability transitions. Springer Japan, Tokyo, pp 13–32

Chapter 3
A Dangerous Transition to Hope

Brendan James Gleeson

Abstract The dawn of a global urban age is the beginning not the end of a challenging journey. Industrial capitalism it seems is finally mired in insuperable contradictions, and transition to a new human dispensation is inevitable and desirable. A two-stage transition awaits: first, a time of uncertainty and painful adjustment as the retrenchment of carbon capitalism begins in earnest, followed, it is to be hoped, by a new political economic order that provides humanity and the biosphere with the means for a safe and sustainable coexistence. My interest in this chapter is with the first stage – of painful adjustment – which will surely necessitate repurposing of state rationale and scope. My controversial submission is that we may need a strong 'Guardian' State to guide the transition and forestall attempts to reinstate the ruinous conditions that have caused the present crisis. In the neoliberal present, this idea is perhaps unthinkable, but neoliberalism is collapsing and the time for radical action has surely arrived. The Guardian State would seek a new human dispensation and represents therefore a postcapitalist response to the dissolution of neoliberalism.

Keywords Climate change • Neoliberalism • Guardian State • Transition • Urban age

3.1 Introduction: Is the Urban Age the Final Age?

We live in what the late Ulrich Beck termed a 'world at risk'. The mounting social and ecological crises that have attended the rise of industrial capitalism during the past three centuries have reached planetary scale in reach and threat. The evidence suggests that a terminal crisis of capitalist modernity is no longer a prospect; it is

---

An earlier version of this chapter appeared in Gleeson (2014).

B.J. Gleeson (✉)
Melbourne Sustainable Society Institute, The University of Melbourne, Melbourne, VIC, Australia
e-mail: brendan.gleeson@unimelb.edu.au

T. Moore et al. (eds.), *Urban Sustainability Transitions*, Theory and Practice of Urban Sustainability Transitions, DOI 10.1007/978-981-10-4792-3_3

breaking upon us. Luminaries of political economy agree. Wolfgang Streeck (2014) asks not if capitalism will end but how. Andre Gorz beat him to it declaring, just before his death in 2007, 'the exit from capitalism has already begun' (2010: 21).

An optimistic counter narrative arises from the chorusing of an urban age. A chorus of expert commentary welcomes a golden era of human prospect (e.g. Glaeser 2011; Brugmann 2009). A new conversation welcomes the fact that humanity is now preponderantly an urban species, homo urbanis. The major transnational institutions (e.g. OECD 2010) bestow great significance to urbanisation as a force shaping human fortunes. For the past half-decade, the United Nations has broadcast the message of a new urban ascendancy. UN-Habitat enthuses, 'A fresh future is taking shape, with urban areas around the world becoming not just the dominant form of habitat for humankind, but also the engine-rooms of human development as a whole' (2012: v). Against this, radical commentary asserts that a globalising world is in fact a morbid *Planet of Slums* (Davis 2006). Planetary urbanisation is explained as the contemporary face of entropic capitalism charting the ruinous courses of over-accumulation, social polarisation and natural destruction (Harvey 2010). Harvey writes:

> . . . so many people in the world live in conditions of abject poverty, that human dignities are everywhere being offended even as the rich are piling up more and more wealth under their command, and that the levers of political, institutional, judicial, military and media power are under such tight but dogmatic political control as to be incapable of doing much more than perpetuating the status quo. (ibid: 228)

3.2 Is the Dawn of the Urban Age the Last, Dwindling Light of Human Prospect?

We moderns are forgiven for fearing that all is lost and that Jane Jacobs' (2004) dire prediction of epochal collapse, a *Dark Age Ahead*, was terribly right. And yet this is not the final truth. We should have faith in the capacity of modern civilisation to survive its greatest failures. Modernity released our species from an aeon of grubbing, toil and servitude. There is no evidence that we humans have collectively given up on this great saga of emancipation. Indeed its human appeal continues to widen in a long, relentless spread across the face of a cosmopolitan globe. It will survive if we do. And we will outlast the present crisis; battered and baked, to be sure, but our species will continue. Even the most appalling scientific predictions (e.g. Lovelock 2009) grant the Earth and its dependants some forms of future. We may have to endure catastrophes, but the cause of catastrophism is wrong headed and certain to deflect us from the task of reclaiming the human prospect from our own follies. Humanity will emerge remade from the storms about to break, as will the planet we inhabit and shape, in ways we cannot yet foresee.

History is replete with moments when civilisations disintegrated after prolonged welling of extrinsic pressure and internal contradiction. This time 'collapse' could mean retrenchment of the entire human order, not a particular civilisation. The end play for capitalist modernity might be a series of rapid acts, a staged breakdown that starts with a rapid worsening of existing social and ecological faults, followed by a generalised dissolution of human and natural systems. The first storms of industrial apocalypse will break in the great and emerging cities of the Global South. It is here that the furies of sea level rise, capricious weather and resource depletion will render first force, doubtless with terrible effect. Wealth and geography may confer a measure of resilience on the historical northern heartlands of industrialism, at least for a time. And yet even here, as Berry (2013) makes clear, decades of neoliberal austerity have bequeathed massive deficits in the resources, including governance capacity, that will be needed to manage the impacts of climate shift.

Collapse might occur at the global scale, and the entire terrestrial order remade. The Earth's climate system, natural resource base, species ecology and terrestrial and aquatic landscapes are now rapidly taking new forms. The rise of a human urban age is part of this wider, planetary-scale process of dissolution and reformation. We may, as many scientists believe (Crutzen and Stoermer 2000), be leaving the Anthropocene – what we now understand as a relatively brief historical period (from around 1800) in which exponential growth of human power and possibility was released by the advent of carboniferous capitalism. We are unprepared for what is to come. Amin states it well: 'If 20th-century modernism clung to the hope of progress for all and mastery over future vicissitude, our times seem to be preparing for a rougher ride, without the confidence of knowing how best to forestall hazard and risk or harness the future for general well-being' (2013a: 140).

Are there resources for hope? Perhaps the greatest insight of modernity was that destiny is choice not fate. The hard work of choice is upon us. Capitalism may well have entered its terminal phase, as the late Andre Gorz firmly believed, but this does not mean the end of humanity or even modernity. Our species will have to take to the road of history again. Looking ahead to prospect, Eagleton writes, 'Hope . . . is what survives the general ruin . . . '. Further, 'Though there will be no utopia, in the sense of a world purged of discord and dissatisfaction, it is sober realism to believe that our condition could be mightily improved. It is not that all will be well, but that all might be well enough.' For the hopeful the ending of the current world order means a new journey not a termination of history. The close of Beck's 1993 book, *The Reinvention of Politics*, ponders the imponderable, a modernity that has run out of ideas, exhausted by hubris and depleted by failures. And yet, he doesn't lose faith in the project, seeing portent in ruin. The last sentence reads, 'Only a final lack of options frees oneself, but you still hope and you're dangerous . . . ' (Beck: 177). Perhaps the most dangerous and hopeful idea in the present is to not fear catastrophe, seeing it instead as means to transformation and renewal.

3.3 Transitions, Management and the Urban Age

There are two great exigencies of our time. There is the work of transformation to the postcapitalist order – to chart and journey towards these shores without delay. But first, or at least simultaneously, a great, possibly prolonged, necessity is to weather the storms of change in the safest harbour we can make for. This is most likely some form of 'war state' at the national level, buttressed and led by a new internationalism that bends global institutions to the cause of human transition and survival. In this chapter, my contribution to 'transitions' discussions is to essay the 'time of storms', to ponder prospects for socio-ecological coordination and governance as the retrenchment of capitalism begins in earnest.

When set against the prospect of the 'time of storms', the 'transition management' literature (e.g. Loorbach 2010) presents an optimistic, reassuring tone, trusting the forces of capitalism to save itself, the planet, and urban humanity from itself, through the intelligent application of levers of policy. Implicit, because barring the emerging work on power introduced in Chap. 1 (e.g. Meadowcroft 2009; Avelino 2009), Transition Management has been largely silent on the political capacity and limits of capitalism to intervene in climate destruction. The suggestions made for governance of transitions typically assume the political economic structure and aim for sociotechnical transitions and not so much transformation of the underlying structures that are producing dangerous climate change. In this, it parallels the optimism of the likes of the aforementioned Glaeser and Brugmann. This chapter starts from the assumption that Transition Management is overly optimistic and speculates from this a more radical, perhaps realistic narrative. Its central, and doubtless controversial, idea is the prospect of a Guardian State to guide and safeguard human possibility during a period of species endangerment. This is explicitly a transitional arrangement intended to guide humanity through the storms of change to hopeful shores. In other work (Gleeson 2014) I have outlined in some detail a new human dispensation that might be strived for during the period of fearful transition.

The chapter begins with a set of short assessments portraying the rapidly worsening course of socio-ecological endangerment that faces the human species. In short the transition is upon us and it is principally in cities that homo urbanis will meet its fate. An outline case for strong progressive governance during the transition is presented. The chapter concludes, as the human journey must, on the question of hope.

3.4 Towards the Storm

We do not yet know how the emergency will unfold, but science is relentlessly bringing it closer to nearer horizons. Authoritative science reports that we will enter a 'new' (and dangerous) world climate before 2050 (Mora et al. 2013). Thus we are entering not approaching a warmer, poorer world. And even as we move through its gates, there will be tireless replay of its destructive origins. The monolith of overproduction remains unchallenged. The carboniferous excess of earlier Western industrialism will continue to flourish in new realms as the crisis unfolds. Contradiction and confusion will cloud human sensibility. In the face of decades of green Western advocacy, Davis reports that:

> Coal production has undergone a dramatic renaissance over the last decade, as nightmares of the 19th century return to haunt the 21st. In China 5 million miners toil under dangerous conditions to extract the dirty mineral that reportedly allows Beijing to open a new coal-fueled power station each week. Coal consumption is also booming in Europe, where 50 new coal-fueled plants are scheduled to open over the next few years, and North America, where 200 plants are planned. (2010: 34)

The necessary goal of decarbonisation will not be realised through reform which must increasingly turn its attention to the project of species survival, rendered by policy and science as the anodyne imperative 'adaptation'. An armada of well-meaning advocacy and action will attempt to steer a way for humanity through the storms. Successful experiments in decarbonising European cities apart, the project of worldwide adaptation to climate change, accommodating tens of millions of climate migrants and a huge redistribution of wealth in the process, would require a revolution of almost mythic magnitude in the redistribution of income and power (Davis 2010: 38).

Well might they try, but the tsunami of human and global change will overtake well-intentioned action. The scale of the problem will continue to magnify both in terms of threat and species exposure. By 2030, when global heating and resource finitude will begin to converge terribly, there will be at least another 1.5 billion humans, mostly in the slums of the Global South.

To speak of species danger is to risk neglecting the differential harm and burden that a collapsing industrial order will impose on the globe's regions and peoples. The poorer nations will bear the greatest early impacts of warming and will be least equipped to deal with them. Even the ebullient optimism of urbanology is blunted by the indisputably terrifying implications of global warming in the South. Kahn (2010), whose *Climatopolis* adopts a remarkably sanguine view of the warming crisis ('. . . our cities will thrive in the hotter future') admits that 'climate change in the developing world will raise the risk of civil war, urban disease epidemics, and mass death from natural disasters' (2010: 158) – with one major exclusion. He is certain that his pin-up model of economic development, China, will use growth and innovation to navigate safe passage through the coming tempests. This contrasts with a rising tide of assessment which sees Chinese urban growth careering towards environmental calamity (Hamlin 2013).

Kahn's anxieties about the developing world are rested by the old biblical injunction about the comforting permanency of the poor. They have 'always been with us' and surely always will. In this long historical light, concern of the day seems hubristic and, in any case, unmindful, not to say disrespectful, of 'their' ability to cope with privation: 'I am concerned about how the urban poor in both rich and poor nations will cope with climate change, but the truth is that this group has always faced hardship' (ibid: 242). Climate response mustn't therefore get too bogged down with such 'group politics'. In any case, and most marvellously, what imperils the poor may actually save them, 'An optimist would say that climate change will create an imperative for nations to embrace pro-growth strategies to help the poor move up the economic ladder' (ibid). Perhaps sensing a slide from optimism to fabulism, Kahn hedges his sunny bet on a 'hotter, better' world, admitting, '. . . much of my optimism goes out the window if climate change inflicts abrupt shocks' (2010: 241). The climate science to which he apparently subscribes forecasts a volatile not smooth transition to a warmer world.

Decades of neoliberal globalism have engendered haphazard and poorly realised urbanisation, bequeathing a legacy of slum-bound cities that are exquisitely exposed to the smallest perturbations of climate, resource flows, or disease. In this sense urban destitution extends beyond its obvious slum expression to describe the utter dearth of defensive and coping resources, and thus resilience, at the city scale. Berry believes that the damage will continue and that austerity governance will '... likely renew neoliberal calls for governments to withdraw further from direct provision and strong regulation' (2013: 9). Crucially the pattern of worsening intra- and interurban inequality that was deeply set within neoliberal urbanism remains unchecked. The relentless drift to neoliberal polarity '... cuts away the moral basis of urbanity, the sense of shared purpose and propos of communal identity' (ibid: 14). Weakened social solidarity has grave material not simply moral consequences, depriving cities of the collective resolve needed to confront climate (and other) threats.

On the frontline of disaster are the low-lying Southern megacities where sea level rise and storm volatility present existential threats. Here Lomborg's (2012) cheerful advice that resilience can be maintained through massive investment in defensive infrastructure (sea walls, flood barriers, etc.) resounds with ghoulish humour. Kahn, the savvy realist, isn't buying it. He offers with simple honesty: '... I would not advise buying property in: Dhaka, Jakarta, Manila and Calcutta' (2010: 186). For him, 'Every shock creates challenges and opportunities', but apparently not everywhere (ibid: 187). Davis fears that '... cities of poverty will almost certainly become the coffins of hope' (2010: 30). His counsel is not complete despair but that 'we must start thinking like Noah' (ibid). This is to echo Lovelock's (2009) insistence that humanity makes fast its 'lifeboat regions'. For homo urbanis, foremost amongst these are lifeboat cities (Gleeson 2010).

Another legacy of global neoliberalism has been the erosion of human solidarity, which now clings to life in faltering, under-resourced institutions such as the United Nations. As Davis and many others point out, the econocratic globals, such as the World Bank, the G20 and the IMF, have undermined rather than strengthened species security and ecological integrity. Global civil society registers a pulse in the NGOs committed to human and ecological flourishing, but none of this amounts to the collective resolve and resources our species will need to endure the storms of epochal change. This, for now, leaves us to the alarming prospect of widescale dissension, and doubtless war, as the crisis unfolds and human populations are uprooted and whole states and societies upended by natural default. Davis asks, '... will the electorates of the wealthy nations shed their current bigotry and walled borders to admit refugees from predicted epicentres of drought and desertification – the Maghreb, Mexico, Ethiopia and Pakistan?' (2010: 38). The defensive, often cruel, chauvinism exhibited by wealthy states towards small floods of asylum seekers in recent years is hardly encouraging.

If our species, and not just its wealthy, should be saved, we must return to Davis' question, Who Will Build the Ark? Perhaps lifeboats offer the better idea, given the plurality and spread of human occupation, including the rising significance of urban concentration. So, when and how? Given that the war we declared upon ourselves has broken out, the arks of human survival will need to be produced like the famous American Liberty ships were in the last great global conflict – rapidly, robustly and with provision for tempest and famine. Importantly, like the Liberty ships, our human arks must be designed for a specific time, that of species transition. They must not become prison hulks for a diminished humanity. These vessels should carry us to newer shores. They will of necessity operate under strictures that must one day be lifted in a new, safer human dispensation.

3.5 Already Dangerous

The transition from industrial modernity, and surely from capitalism in its present form, is underway – an overdetermination of human prospect by shifts that are both contradictory and mutually reinforcing in myriad ways. They include all major dynamics of social, ecological and economic change which have now adopted common transformative qualities. Stasis and equilibrium are footnotes in neoclassical imaginary and not present to the contemporary human condition. We have entered as Amin puts it '. . . a world in disequilibrium' (2013a: 140). The greatest force for transformation – surely the most potent ever known – is a warming global climate.

As noted earlier, a 'safe climate' is now behind us, not around us. Prospects for return to late Holocene carbon dioxide levels (in the range 300–325 ppm) are as yet unknowable. Spratt's (2013) review and summary of core science and of leading institutional opinion map the contours of danger. As Spratt concludes, the arrival of dangerous warming, dynamically fed by 'positive feedbacks' in the wider climatological system, leaves humanity with two stark options, both of which mean abandoning the fantasy of a two-degree stabilisation scenario. He states, 'The real choice now is to try and keep the planet under a series of big tipping points by getting it back to a Holocene-like state, or accept that a 3–6 °C 'catastrophe' is at hand' (ibid). Two transition paths thus lie before us – to a post-Anthropocene world in which human possibility is radically diminished but not extinguished or return to something like we have now climatologically but having discarded carboniferous capitalism. The latter means climate recovery but not the reinstatement of the order that destroyed it in the first place. It signals, as Heinberg explains, a new Anthropocene founded on '. . . a steady state-economy based on a recognition of ecological limits' (2009: 167). Given the warming already 'locked in' by past emissions, we will have to make great strides – unprecedented as a species – to regain the planetary ecology of the present.

The first path must be refused, but to take the second requires deep subscription at the species scale to what Spratt terms 'a radical future'. Heinberg (2009) argues that we should deploy these last high net energy 'riches' to fund the work of transition. Yet the massive, sudden cuts in emissions it dictates cannot surely be achieved without first discarding faith in the mitigation mechanisms presently favoured (economic instruments, technology, green behaviourism) yet so far patently insufficient to the task. A radical shift beyond the tools of Transition Management is needed in the form of wholesale political economic change to shift the structures that produce and sustain carboniferous capitalism. This raises the prospect of the suspension of capitalism, as in previous world wars – and its replacement by an entirely new mode of production that maintains human impact within safe ecological limits. A radical transition pathway is as follows: first, decisive emergency measures to avert climate catastrophe and restore the planetary ecology and, second, the realisation of a new political economy and, as part of this, a new political ecology that finally lifts the long curse of prometheanism.

3.6 The War on Climate Terror

Spratt reports growing expert and scientific sentiment for a 'wartime mobilisation' to realise the radical goal of climate safety. Brown's (2009) *Plan B 4.0* makes this case with verve. Stretton (2005) has already observed that wartime resource rationing will be needed to rein in harmful consumption. One of the dark natalities to emerge in this path is that democracy, greatest of modern political virtues, may be inhibited temporarily to ensure its survival. The anaemic liberal democracy bequeathed by decades of neoliberalism might be supplanted by nation states on war footings, joined to common purpose by an emboldened and reauthorised United Nations. Pusey (2008) sees opportunity here as climate change-induced political change might restore the legitimacy of state intervention and generate the needed cultural energy for nation-building government.

The dangers, however, are obvious. Waiting in the wings are expert and technocratic rationalities, including the new urban positivism, which may counsel the extinction not careful circumscription of democracy. The cause of species authority might be sidelined by eruptions of authoritarianism. In the latter scenario, the chance to use the transition to forge a new human dispensation would almost certainly be lost and the worst depredations of lethal modernity revisited upon an already weakened and dispirited species. Urry (2011) fears that sudden catastrophic events – collapses in resource stocks (especially oil or water) or climate-related urban disasters – will catalyse the sudden rise of 'war regimes' that dispense altogether with democracy and solidarity in desperate quest for survival of the fittest and favoured. Resort to dystopian warlordism is surely plausible if species action is delayed to the point where it is catastrophes not human will that provoke response. It is urgent that the terms of transitional authority be defined and instated before raw power is given a chance to determine them. To drift further into the emergency without beginning a democratic response will be to give way to the claims of supremacy.

This dark possibility encourages the need for new global governance for human survival to be forged now, not in the teeth of the worst storm furies. When the storms begin to break from around 2030, the vulnerable cities and hinterlands of the Global South will house at least 80% of the world's urban population: much of their fabric will be woefully unshielded slums. Amin (2013a) is impatient with academic and activist 'revisionism' that finds wondrous adaptive resources within 'informal' (i.e. slum) urbanisation. As he points out, cities contain immanent potential for endurance in the face of external threat, through capacity for social mobilisation and deployment of 'machinic infrastructure'. But such possibility has been radically diminished by prolonged neoliberal urbanisation, especially in the South (Berry 2013). A more fundamental, planned response from reformed global institutions is needed to repair the defences of developing cities and begin transition to a new model of resilient urbanisation. This presupposes a massive programme of investment by the Global North in a wholesale strengthening of the world urban system, beginning with its most susceptible parts. The case for this historic transfer of species wealth has been well essayed and retailed (e.g. Brown and Eriksen 2011). Indeed the mechanism exists in incipient form. The Climate Adaptation Fund set up under the Kyoto Protocol of the UNFCCC is a modest and halting start on this work, but proof that the concept could potentially gain wide human assent.

What of governance in an urban age of dangerous transition? All our ingenuity and resourcefulness must be immediately deployed to the task of fashioning a new urbanism that will, in order of priority, first, bring us through the tempests of transition and, second, provide the new settlement pattern that a post-carbon future will necessitate. Presently, and morbidly, much human urban innovation is directed away from this end, towards the quicksands of the compact city or the absurd or elitist stage sets of corporate ambition. Emblematic of the latter is Facebook's new company village, Anton Menlo, which replaces the progressive idealism of the historical model town with the privilege reserved for the knowledge worker of neoliberalism (Albergotti 2013). As related earlier, a new progressive urban imaginary is needed, especially directed at defining city life and functioning in a new postcapitalist modernity that commits to safe co-evolution of humanity and nature. Without diminishing the need for this act of imaginative and material creation, a new model of transitional urban governance must be agreed and implemented immediately to manage dissolution of the 'second modernity', the unstable risk-borne age that has largely replaced industrialism and which is now itself breaking apart (Beck 2009). Transitional governance is not only about survival as an end in itself but to begin the identification of values and structures that will, when conditions allow, eventually create a newly stabilised 'third modernity'.

3.7 A Guardian State

If we accept together with many others that climate change is unmanageable under capitalism, then it follows that recovery from climate change and response to it will require an end to modern capitalism, in a radical and contingent transition.

The Guardian State is a postcapitalist response. It is presented here as a 'logical' possibility along with speculative 'transition' ideas that might be used to bring policy/governance levers to bear in a mitigative way. The following sketch intentionally leaves aside questions about processes of transition and instead focuses on the central concept of crisis management in the war on climate change. While it brings obvious dangers, it also presents a possibility through which neoliberal urbanism would be swept away by a new tide of human assertion. In this prospect, the Predator State (Galbraith 2009) that has protected and advanced the (il)logic of laissez-faire in the face of contradiction and opposition would give way to a Guardian State that commits to collective welfare in a time of shared endangerment. The central argument is that, as Gorz puts it, 'The exit from capitalism will happen . . . one way or another, in either a civilized or barbarous fashion' (2010: 27).

Preemptive transition to a Guardian State is prospected in such a way that avoids the dystopian consequences of waiting for the inevitable and possibly violent end of neoliberalism and the aforementioned scourge of dystopian warlordism. This state of progressive transition would guide humanity through a time when 'normal' social and economic systems are suspended, some perhaps abolished. It would enable humans to arrest the lethal monolith of overproduction and replace it with social coordination of resource allocation as means both to cope with immediate crisis and to begin transition to a new political economic order. Central to this break with neoliberal logic is to dispense with faith in markets, technology and 'behavioural adjustment' as means to reset production and consumption within safe environmental and material limits. Resource usage reduction for sustainable development sits alongside wholesale decarbonisation to prevent catastrophic climate change. Gorz states the case for a war footing that enables the necessary rationing:

> De-growth is . . . imperative for our survival. But it presupposes a different economy, a different lifestyle, a different civilization and different social relations. In the absence of these, collapse could be avoided only through restrictions, rationing and the kind of authoritarian resource –allocation typical of a war economy. (2010:27, original emphasis)

Amin depicts this as urban ideal:

> This is the city of modest consumption, shared goods and services, alternative technologies, reduced waste and recycling, taxed excesses, pooled technologies, public transportation, and standard housing, all defended as environmentally urgent. (2013b, Fn3: 204)

Such a state would embody a level of resolve that liberal democratic governments cannot possess. Equally, however, it must be conceived as a project for emergency times, not for the future. Its foundational premise must be limited to guiding our species through epochal change to a third modernity where a heightened centralisation of state power will not be necessary. The guiding values of state guardianship must be the 'lifeboat ethics' that Lovelock (2009) speaks of: a trinity of social values, restraint, sacrifice and solidarity. Importantly, the creation of this state means the suspension of capitalism, at least as we have known it. With this comes the necessity of greatly curtailing the power of capital and its servant lobbies and heralds, especially those in the rearguard of carboniferous capitalism delaying transition from resource- and energy-intensive economies. A gift of the crisis should be to terminate the basis for accumulation and exploitation entirely.

In a world in thrall to austerity governance and its beneficiaries, the suspension and/or restraint of private power might seem like impossibility. The difficulty of subjecting private, especially corporate, prerogative to social priority is formidable. It was just so in World War II, when Allied states suspended markets and many other things besides, including corporate entitlement, property rights and some civil rights. Wholesale freezing of 'the system' was democratically possible. The sense of war emergency was reinforced by the popular disappointment in its functioning generated by the Great Depression. Progressive politics must work to rechannel the human disenchantment that attends the present dissolution towards a new species will, guided by collective not individual values. The case for radical intervention must be made and carried with urgency.

How will democracy be firstly restrained, yet somehow protected, and then subsequently restored? The short answer is that politics and the state might function something like they did in the Allied world during World War II. A multilateral consensus authorises strong central coordination for a time to address a national and global emergency. Libertarians and communitarians will argue that power should be devolved to households, communities and individuals who could find a Lilliputian way through the crisis via 'well-informed' actions. This is a dangerously wrong-headed dream. Civil rights would need to be preserved by the Guardian State, but this does not mean recourse to libertarianism. Nor can it sanction continued institutional tolerance for the scepticisms and refusals that deny the basic terms of the crisis and thus undermine the collective resolve needed to meet it. The Guardian State idea is fraught with such concerns, yet what are the alternatives when faced with the transition imperative for human survival?

The period of emergency would require restraint of many habits and practices, largely to do with consumption and production that would harm the collective good in a time of stress. Civil rights are preserved and consumption 'rights' not. It might, for example, be necessary to restrain forms of energy use (e.g. air conditioning, air travel), but measures like this will not interfere with cultural and political expression. Such a stressful period would require the transparent enactment and display of fair treatment for all, as never before. The divisive legacy of neoliberalism would need to be swept aside. Immediate introduction of a guaranteed subsistence income would commence the labour of repairing social solidarity (Gorz 2010). Maintaining, even strengthening civil rights, would be part of this. The Guardian State would need to carry the fairness ideal on its breastplate as we enter the battle for human recovery.

The State should also begin to restore the most important and yet severely degraded democratic resource, the public sphere. In the wake of neoliberalism, this state would do more to restore democracy than to restrain it. To rebuild the public sphere – that of civic praxis – would set course for the Arendtian ideal of a fully realised human urban society. In her vision, it is the polis committed to political action – that is, the discursive, imaginative essaying of the possible – which nurtures the immanent human gift of natality. It is the city of political action that gestates and brings forth the 'miraculous discoveries' that defy the odds presented by naturalism. By these means humans make themselves and their works 'imperishable' even in the face of dire threat (Arendt 1998[1958]: 197–8). The crisis will shatter the post-political neoliberal city which is the antithesis of this ideal.

3.8 Lifeboat Cities

State rescaling is needed to strengthen governance of the world's cities which have heretofore been deprived of the means for effective coordination by the diktat of neoliberal urbanism. Metropolitan governments with clearly scaled authority must guide humanity through the fierce dawn of the urban age. Amin underscores the weakness of contemporary metropolitan and local governance and thus the magnitude of effort needed to gird them for the task of urban transition:

> In most countries, [they] remain hampered by juridical and functional restrictions and are frequently captured by vested interests or held back by lack of resource, capability, commitment, and efficiency, while national governments – with far greater powers and resources – remain blind to urban centrality. (2013b: 203–4)

Transition to a guardian metropolitan state model means a new and recently unprecedented commitment to social coordination of the urban process. Amin provides the written orders for the new agent of urban leadership:

> ...tackling powerful nongovernmental forces shaping urban outcomes: the command centers of global supply chains and business and financial transactions, the social networks of urban elites, the spaces of knowledge production and cultural validation, the intelligence of hidden technological infrastructures, and the transactions of the informal and illegal city. (ibid.)

Metropolitan governance through the crisis obviously necessitates a considerable degree of authorisation and autonomy, but carefully and explicitly coded to prevent drift to autarky and defensive resilience. While nation states will retain authority for setting the terms of transition at the country level and for supporting the new internationalism that will, inter alia, secure the safety of the Global South, urban isolationism will need to be somehow insistently prevented by multilevel governance that inscribes commitment to human solidarity as a leading value.

Cities must be reconceived as the lifeboats that will carry a chastened species through to safer shores. Human assertion and collective confidence must be brought back to the fore. In this sense, there is something to be admired and learned from the bravura of urbanology and the confidence of urban physics. They may be weighed down by faulty assumptions and bad science, but the idea that cities could be the basis for a revolution of human prospect (to restate Glaeser) makes compelling sense. Within the sunny field of urbanology, there is partial recognition of the need for decisive action, not simply reform, to avert climate disaster. Hollis upbraids

Kahn for his faith in market adaptation, observing dryly, 'We are going to have to make rather more concrete plans for the future' (2013: 305).

Importantly, urban compaction cannot be seen as panacea. Its physical determinism shrouds the need for massive social coordination of consumption and production to stave off danger and rescale human environmental demands within safe limits. It also underestimates the risks of high-density urban forms in many contexts and the potential (if latent) adaptive capacity of suburbia. Planning would be freed from the deflecting ambition of frontline mitigation and reauthorised to achieve resilience as a key urban strategy of the Guardian State. Urban policy (transport, planning, building, design) must, of course, also take on the longer, larger task of recreation, to produce built environments that use vastly less energy and resources than presently.

In the wake of neoliberalism and its many denudations of collective rationale, the Guardian State may seem like a life raft kit without instructions. And yet many of its values and practices are readily seen in historical examples of war time conditions. In urban settings, Barry (2012) finds the ideological and practical bases for collective action sketched in progressive collaborations of citizenry, such as the Transition Towns Movement and its many kindred networks and alliances (see Hopkins 2008). Such mobilisations (Gorz (2010: 126) calls them '...persuasive forms of social experimentation') suggest that transition is already underway, albeit at a small scale, perhaps awaiting leadership and authorisation by a state committed to a fundamental realignment of social purpose. They reveal for wider consideration and take up, '... 'concrete utopian' practices in the sense of being experiments within a self-transforming present as opposed to the creation of some new future based on abstract principles' (Barry 2012: 115). These social experiments exist despite contradictions with capitalist practices of accumulation. Imagine how transition practices might be brought to bear to accelerate such experiments in a support of a war effort against climate terror, as part of a Guardian State initiative that effects a shift from such dispersed, 'niche' grassroots collaborations to a broadly coordinated movement.

We should believe that a place of safety if not tranquillity lies beyond the storms of the present crisis. This is a destination that the species itself must choose not anticipate, through an enormous and unprecedented act of collective will. Massive decarbonisation and resource restraint are necessary steps towards a safer place. They will be all the harder to achieve as we suffer the depredations of a failing world order. Retrenchment of capitalism will not present a straight path back to ecological moderation. Its death agonies will likely generate many wild quests for salvation through vulgar resource exploitation. These deathly misadventures are prefigured in the contemporary lust for Arctic exploitation, the fracking rush in the new worlds, and the enthusiasm for newly unlocked carbon, such as Canada's tar sands. The Guardian State presents as a means not only to secure human populations in the storm but to stymie the suicidal urges of a dying order. This means, inter alia, forbidding piratical raids on residual nature.

3.9 Conclusions: Imaginaries of Hope

In a time of fiercely fought survival when species wit must be turned to many, rapid fire material dilemmas, we must also reawaken the human imagination. At some point the terrible ash cloud of neoliberalism will lift from human thought. If we are not to remain trapped, in its aftermath, like the lifeless clay forms of Pompeii, there must be some new animating sense of purpose and direction. Humanity will need to assert, indeed arrogate, a great expedition for safety and realisation. We need to do this now, not later, 'after the fire' of crisis, because species direction, like a supertanker, is a long steady course best set at the first opportunity. Davis states it thus: '. . . only a return to explicitly utopian thinking can clarify the minimal conditions for the preservation of human solidarity in face of convergent planetary crises' (2010: 45). In a world of intellectual quietude, these words defy the silencing order of neoliberalism.

What guiding stars can we discern, now that the heavenly dome of power is fracturing? This is the exquisite dilemma of our dangerous age. Harvey puts it plainly, 'Our political task . . . is to imagine and reconstitute a totally different kind of city out of the disgusting mess of a globalizing, urbanizing capital run amok' (2012: xvi). A post-political order and dimmed social science mean human discernment is at low ebb. As Arendt made clear, however, the human condition remains open to the birth of unforeseen possibilities. In a time of planetary dissolution, we have never needed our fabled imaginative powers as much as we do now. In this sense, the urgent task amounts to greatly expanding the horizon of existing Transition Management agendas to face the political project of transition.

References

Albergotti R (2013) 'Facebook's company town', Wall Street J, 3 October, Accessed at: http://online.wsj.com/news/articles/SB10001424052702303492504579111792834660448, on 10 Oct 2013

Amin A (2013a) Surviving the turbulent future. Environ Plan D: Society and Space 31(1):140–156

Amin A (2013b) The urban condition: a challenge to social science. Publ Cult 25(2):201–208

Arendt H (1998[1958]) The human condition. The University of Chicago Press, Chicago

Barry J (2012) The politics of actually existing unsustainability. Oxford University Press, Oxford

Beck U (1993) The reinvention of politics. Polity, Cambridge

Beck U (2009) World at risk. Polity, Cambridge

Berry M (2013) Neoliberalism and the city: or the failure of market fundamentalism. Housing Theory Soc 31(1):1–18

Brown L (2009) Plan B 4.0: mobilizing to save civilization. W. W. Norton & Company, New York

Brown K, Eriksen S (2011) Sustainable adaptation to climate change: prioritising social equity and environmental integrity. Routledge, London

Brugmann J (2009) Welcome to the urban revolution. University of Queensland Press, St. Lucia (also Bloomsbury Press USA)

Crutzen PJ, Stoermer EF (2000) The "Anthropocene". Glob Chang News Lett 41:17–18

Davis M (2006) Planet of slums. Verso, London

Heinberg R (2009) Blackout: coal, climate and the last energy crisis. New Society, Gabriola Island
Davis M (2010) Who will build the ark? New Left Rev, 61 (January–February):29–46
Galbraith JK (2009) The predator state: how conservatives abandoned the free market and why liberals should too. Free Press, New York
Glaeser E (2011) The triumph of the city. Penguin, Harmondsworth
Gleeson BJ (2010) Lifeboat cities. UNSW Press, Sydney
Gleeson BJ (2014) The urban condition. Routledge, London
Gorz A (2010) Ecologica. Seagull Books, London
Hamlin K (2013) Soviet-style suburbia heralds environmental disaster. Aust Financ Rev 8(November):36
Harvey D (2010) The enigma of capital. Profile Books, London
Harvey D (2012) Rebel cities. Verso, London
Hollis L (2013) Cities are good for you: the genius of the metropolis. Bloomsbury Press, London
Hopkins R (2008) The transition handbook: from oil dependency to local resilience. Green Books, Cambridge
Jacobs J (2004) Dark age ahead. Random House, New York
Kahn ME (2010) Climatopolis. Basic Books, New York
Lomborg B (2012) Simple solutions to superstorm sandy. The Australian, 19 November, Accessed at: http://www.theaustralian.com.au/national-affairs/opinion/simple-solutions-to-superstorm-sandy/story-e6frgd0x-1226519124440, on 23 Aug 2013
Loorbach D (2010) Transition management for sustainable development: a prescriptive, complexity-based governance framework. Govern Int J Policy Admin Inst 23(1):161–183
Lovelock J (2009) The vanishing face of Gaia: a final warning. Penguin, London
Mora C, Frazier A, Longman R, Dacks R, Walton M, Tong E, Sanchez J, Kaiser L, Stender Y, Anderson J, Ambrosino C, Fernandez-Silva I, Giuseffi L, Giambelluca T (2013) The projected timing of climate departure from recent variability. Nature 502:183–187
OECD (2010) Cities and climate change. OECD Publishing, Paris
Pusey M (2008) In the wake of economic reform . . . new prospects for nation-building? In: Butcher J (ed) Australia under construction. Nation-building – past, present and future. ANU E-Press, Canberra, pp 17–31
Spratt D (2013) Is climate change already dangerous? Climate Code Red, Melbourne, Accessed at: http://www.scribd.com/doc/168483927/Already-Dangerous-1, on 27 Sept 2013
Streeck W (2014) How will capitalism end? New Left Rev 87:35–64
Stretton H (2005) Australia fair. New South Press, Sydney
UN-Habitat (2012) State of the world's cities 2012/2013. UN Human Settlements Programme, Nairobi
Urry J (2011) Climate change and society. Polity, Cambridge

Part II
Governance of Urban Transitions

Part II presents four chapters which look at different aspects of urban transitions governance through four different case studies. Chapter 4 discusses the requirement for a new approach to strategic policy development with key sociotechnical transitions principles forwarded to reimagine and re-empower the practice of strategic spatial planning, better positioned to address future sustainability challenges. This is followed by Chap. 5 which investigates if the provision of targeted information on theories of sustainability transitions could strengthen organisational strategic planning. Chapter 6 extends the concept of 'boundary organisation' to include other boundaries such as jurisdictional ones in addition to the science-policy boundary and discusses how such organisations might support innovation and enable transitions in public policy institutions and governance. Concluding this section of the book, Chap. 7 looks at what makes some niches more successful than others, both in terms of their establishment and in influencing the broader regime.

Chapter 4
Strategic Spatial Planning and Urban Transition: Revaluing Planning and Locating Sustainability Trajectories

John E. Morrissey, Susie Moloney, and Trivess Moore

Abstract Despite often stated sustainability goals, much of traditional planning practice remains concerned with facilitating the market and maintaining the status quo rather than challenging and transforming it. In this chapter, the planning system is the focus of a sociotechnical systems perspective analysis. This chapter examines strategic spatial planning at regional and city scales through the lens of sociotechnical transitions concepts to provide insight into the role and capacity of spatial plans and planning processes to challenge the status quo and achieve sustainable urban transitions. We present two cases of strategic planning during the first decade of the 2000s at a national and metropolitan scale in Ireland and Melbourne (Australia), respectively – two cases where strategic spatial plans aimed to achieve sustainable land-use outcomes but where planning failed to act as a brake on booming housing markets and related urban sprawl. This chapter also reflects on the lessons from spatial planning processes to inform sociotechnical systems research pointing to the need to incorporate conceptualisations of space, place and context-specific governance in problem framing particularly in considering the challenges of long-term sustainable land-use transitions. We query whether the prevailing planning system common in most developed contexts can be treated as a stable regime, and if so, what benefit this perspective may provide to planning practitioners.

J.E. Morrissey (✉)
Geography, School of Natural Sciences and Psychology, Liverpool John Moores University, Liverpool, L3 3AF, UK
e-mail: j.e.morrissey@ljmu.ac.uk

S. Moloney
Centre for Urban Research, School of Global, Urban and Social Studies, RMIT University, Melbourne, Australia
e-mail: susie.moloney@rmit.edu.au

T. Moore
Sustainable Building Innovation Laboratory, School of Property, Construction and Project Management, RMIT University, Melbourne, Australia
e-mail: trivess.moore@rmit.edu.au

T. Moore et al. (eds.), *Urban Sustainability Transitions*, Theory and Practice of Urban Sustainability Transitions, DOI 10.1007/978-981-10-4792-3_4

Keywords Ireland • Market-led development • Melbourne • National Spatial Strategy • Strategic spatial planning • Sustainability • Sociotechnical transitions

4.1 Introduction

4.1.1 Background

The urban planning system incorporates key mechanisms and processes to direct and shape city evolution including sustainability outcomes. The increasing integration of sustainability concepts into planning policies and practices over recent years potentially constitutes a new purpose, legitimacy and authority for the discipline and its practitioners. However, measures to shift urban and regional development to sustainable, resilient and low-carbon directions remain elusive outside of ad hoc, small-scale interventions. Planning policies and practices broadly continue to maintain the status quo of unsustainable 'business-as-usual' development models such as dispersed low-density car-based housing with inadequate sustainable transportation. In most developed countries, planning has undergone a process of neoliberalisation over several decades, dated from after 1970 by Gleeson and Low (2000) and described as acquiring 'a central position' since the fall of communism by Sager (2011, p. 149). The result is that planning as a function of the state acts primarily as a facilitator of economic development, as opposed to a means of robustly guiding, managing and coordinating urban development to achieve a range of economic, environmental and equity goals. The dominance of the market-economy paradigm which prioritises short-term economic goals represents a significant structural barrier. Gleeson and Low (2000, p. 96) describe the new neoliberal approach to governance as based on the 'ideological trilogy of competition, deregulation and privatisation'. The frequent partial or total suspension, amendment and revision of spatial plans in response to market priorities serves to undermine the often legitimate strategic work invested in developing long-term plans. A lack of coherence in planning documents, inefficiencies and gaps in implementation mechanisms often opens up room for discretion in decision-making, fuelling uncertainties and undermining sustainability goals.

This chapter examines strategic spatial planning at regional and city scales. The potential of sociotechnical transitions concepts to provide insight into, and understanding of, inherent challenges with strategic spatial planning as currently practised is assessed. Two cases of strategic spatial planning are critically reviewed. First, we draw on the urban planning and sociotechnical transitions literature to frame the role of urban planning as a process for enabling sustainability transitions within a multifaceted and complex urban sociotechnical regime. Second, we present two cases of strategic urban plans developed in the 2000s at a national and metropolitan scale in Ireland and Melbourne, respectively, as examples of long-term strategies emerging from two different planning systems; while including visions and goals for sustainable urban development, both systems failed to guide successful outcomes. Thirdly, we discuss, in both cases, the gap between strategic and operational capacity within existing planning processes, highlighting the obduracy of regimes governing urban development, and draw insights from Transition Management and strategic niche management in understanding the capacity of emerging niche innovations to reconfigure land-use regimes and enable urban sustainability transitions.

4.1.2 Planning for Sustainability

Much of traditional planning practice remains concerned with maintaining the existing social order rather than challenging and transforming it (Albrechts 2015). In many developed countries, neoliberal policies have resulted in increasing critique of progressive planning, ranging from arguments against spatial planning to positions that accept some form of intervention in certain limited contexts (Eraydm 2012). Spatial planning has been undermined on two interrelated fronts as a result of neoliberalism: from the economic and financial system that requires fewer constraints on the reproduction of capital and from a weakening of the regulatory capacity of the state over land use (Goncalves and Antunes Ferreira 2015). Since the 1980s the planning system has been gradually 'neoliberalised' to be highly facilitative of market rather than societal interests (Murphy and Fox-Rogers 2015).

According to Goncalves and Antunes Ferreira (2015), the 'neoliberal turn' has reinvented planning with a focus on 'competitiveness', whereby planners and local authorities adopt a more proactive and entrepreneurial approach aimed at identifying market opportunities and assisting private investors to take advantage of them (Albrechts 2015). This competitiveness agenda has been widespread across much of the developed world. Within a neoliberal frame, spatial planning may be opposed as excessive public interference, resulting in limited scope and influence of the planning process in certain regions (Thornbush et al. 2013). As a result, planning as a guidance, management and coordination mechanism between the economic and social spheres has become fundamentally questioned (Gunder 2006). Nevertheless, spatial planning remains a critical tool for city-region decision-makers to address challenges such as climate change mitigation and adaptation (Kumar and Geneletti 2015). Spatial plans guide the physical development of the city, determining the configuration of infrastructures like transportation and housing as well as resource allocations for social and public amenities (Kumar and Geneletti 2015). Local-level spatial plans are the basic policy framework that shapes social, economic and physical development of cities, while interaction of these aspects is a key driver of climate change (Kumar and Geneletti 2015; Wilson and Piper 2010). In this context, the city region represents a key arena and scale for broader decision-making for sustainability (Jabareen 2015). Birkmann et al. (2014) argue that climate change adaptation and strategic planning discourses should be linked. Despite this necessity, planning to achieve social equity and environmental sustainability goals remains 'stymied by the hegemony of neoliberalism that narrowly frames planning as the facilitator of economic growth' (Rydin 2013 in Leshinsky and Legacy 2015: p. 2). Many global city plans ignore the issue of climate change with a priority instead for economic growth.

4.1.3 Analytical Framework and Purpose of this Chapter

The rapidly growing transitions literature presents a wealth of conceptual richness and depth through which to interrogate planning and its relationship to sustainable outcomes in urban settings. Authors such as Murphy (2015), Hansen and Coenen (2015), Bridge et al. (2013) and Raven et al. (2012) discuss the spatial aspects of transitions, for instance, and point to a dearth of spatially conscious transitions-related study. In this volume, de Haan's (see Chap. 2) concepts of *locus* and *nexus* point to intriguing new directions for planning-related research. There is a wealth of transitions research on governance (e.g. Kern and Rogge 2016), politics (e.g. Meadowcroft 2011), stakeholders (e.g. Gazheli et al. 2015) and policy networks (e.g. Normann 2017). While of clear relevance to this study, a fuller discussion of this large body of literature is beyond the scope of this chapter. In this chapter, we apply understanding from two strands of transitions thinking in particular, the Multilevel Perspective (MLP) and Transition Management (TM) approaches. The rationale for applying these two transitions approaches is as follows: The MLP provides contextualisation of embedded barriers to sustainability within wider social and economic systems, as well as providing a useful heuristic for framing niche innovation and landscape dynamics. In this paper, MLP is applied to provide a broad framing for interrogating processes of inertia, innovation and landscape dynamics within planning systems. Figure 4.1 provides an overview of the MLP, as conceptualised in this chapter.

A central tenet in MLP is the stabilising influence of the sociotechnical regime, (Coenen et al. 2012). Within the urban sociotechnical regime, a number of constituent, linked or 'nested' systems can be identified, including energy, transportation and land use systems. In this chapter, we conceptualise spatial planning systems as a subset of a wider urban governance system, itself a key component of a larger urban sociotechnical regime (Fig. 4.1). We posit that while the broader urban sociotechnical regime itself forms a stable structure and relatively fixed development trajectory, the elements of the regime can be dynamic within this, evidenced by the changing role of planning. Figure 4.1 shows subsystems of the urban sociotechnical regime, which at given times may be in contest, in cooperation or in tension with each other. Within this framing, we investigate spatial planning practice as a sociotechnical process within the urban sociotechnical regime. While Fig. 4.1 positions spatial planning as a central, linking and determining subsystem of the urban regime, we question the role and capacity of the spatial planning system in the context of future transition. For this, we draw on TM principles as a means to reflect on strategic spatial planning processes in our two contexts. TM is a form of

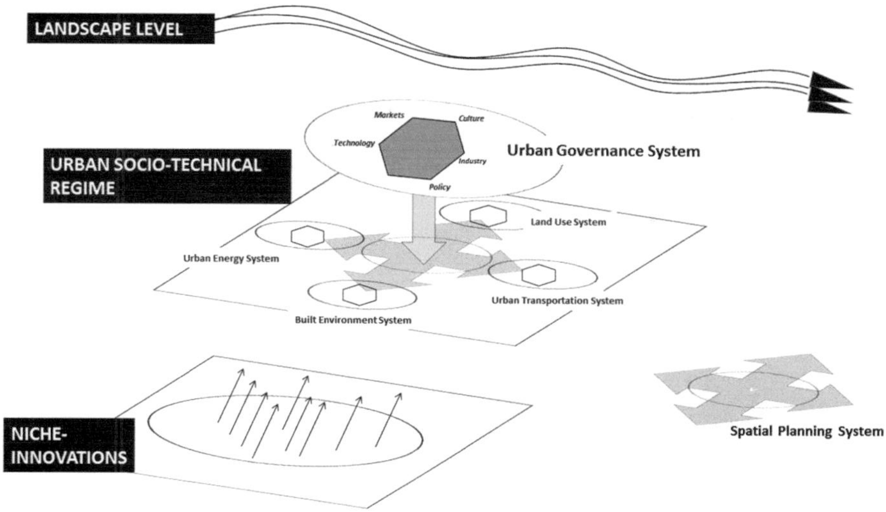

Fig. 4.1 Spatial planning system, linked with 'nested' systems of the urban sociotechnical regime within the MLP (Diagram developed after schematic by Geels 2012)

reflexive governance which means it is understood as part of the dynamics which are governed (Kemp and Loorbach 2006).

We identify processes and characteristics lacking in current planning processes and what might be needed to enable urban sustainability transitioning through the work of strategic planning. Key elements of an applied transitions approach include (Kemp and Rotmans 2009; Rotmans and Loorbach 2008):

- Long-term thinking and setting of visions and goals, which informs short-term policy development
- Integrated policy: multiple domains, actors and levels, links to wider national and international policy development
- Innovation space: the establishment of a transitions arena for technology and social innovation, programme development and ongoing learning
- Innovation policy support: policy oriented towards system innovation besides system improvement (deep structural changes)
- Reflexive governance: to ensure that the transition is 'on track' and avoid a lock-in of technologies and practices
- Multi-actor approach: identification and engagement of societal actors

Applying these transitions concepts, three research questions are addressed:

1. *How can planning practice be understood as a sociotechnical process and element of a broader urban sociotechnical regime?*
2. *What does TM tell us about steering such processes?*
3. *What insights can sociotechnical transitions concepts provide to the revaluing of planning as a tool for change?*

4.2 Spatial Planning Case Studies

4.2.1 Selection of Case Studies

To answer the research questions, two case studies of strategic spatial planning, one in Australia and one in Ireland, are critically reviewed through a sociotechnical transitions lens. The case studies are selected based on a number of parallels and commonalities: both case studies are characterised by market-driven approaches and experience of neoliberalism in planning; the studied spatial plans cover similar timeframes; in both cases, there is common recent experience of housing booms, with associated urban sprawl and negative environmental outcomes; the studied plans display a similar trajectory of state-of-the-art strategic plan development and subsequent marginalisation of these at the implementation stage. While, spatially, the scale differs somewhat in absolute terms, both studied plans are focused on similar population sizes (3–5 million). Finally, while these commonalities were identified as being central to the study, it was deemed that case studies from different jurisdictions would enable distinction between significant local factors and common urban sociotechnical regime characteristics. As will be argued, the large-scale failure of strategic planning policy to direct development to more sustainable outcomes in the face of rapidly expanding economies and associated housing demand shows remarkable consistency and a similar trajectory over the 2000s, in both the Melbourne city region and in Ireland, particularly in the Greater Dublin Area. In this chapter, we investigate how transitions concepts can help to expose systemic shortcoming experiences in both cases.

4.2.2 The National Scale: Strategic Spatial Planning in Ireland

4.2.2.1 Context for Strategic Spatial Planning

Significant interregional differences in economic performance are evident in the Republic of Ireland, including regional disparities in physical infrastructure, human capital and institutional capacity, and related differences in per capita productivity levels (Walsh 2013). Dublin is the state's only metropolitan city region and a distinct lack of large cities elsewhere in the state has negative implications for population growth, economic development and future development patterns (Hughes 2014). Underperforming regional economies have their root in historical policy decisions. The urban-centric Buchanan Report (1968) advocating a national city-growth focus was firmly rejected by the government of the day, in favour of a more populist 'scattergun' approach, whereby industrial branch plants were located across a range of regional towns (Hughes and Sirr 2015). The impact was that Irish cities did not grow as a result of concentrated investment and development (Hughes and Sirr 2015).

The publication in 2002 of the *National Spatial Strategy for Ireland 2002–2020* (NSS) provided a response to the growing imbalances in socioeconomic development, increasingly evident during the Celtic Tiger period of the late 1990s (Meredith and van Egeraat 2013). The NSS was formulated against a background of unprecedented economic transformation in the Republic of Ireland; over a relatively short period, the country had moved from having low levels of economic development relative to European peers to amongst the highest (Walsh 2013). In the aftermath of the Northern Irish Peace Process, it was timely to consider the spatial dimensions of development, with a view to all-island cooperation (Walsh 2013). The NSS represented a departure from conventional planning in Ireland by taking a more holistic perspective of changing geographies of population, settlement patterns and the distribution of employment opportunities (Meredith and van Egeraat 2013).

4.2.2.2 Plan Proposals Versus Actual Outcomes

While the creation of statutory regional authorities in 1994 provided a first impetus for the development of a regional perspective to governance in the Irish state, these authorities were weak and underfunded (Walsh 2014). The NSS was mandated to identify broad spatial development patterns for areas and establish indicative policies in relation to the location of industrial development, residential development, service provision, rural development, tourism and heritage (NSS 2002). To achieve this, the NSS proposed the classification of 18 cities and towns and their associated hinterlands as ‘gateways’ and ‘hubs’, locations conceptualised as being interlinked by social, cultural and, primarily, economic activity patterns (Meredith and van Egeraat 2013). These centres were to provide economies of scale and critical mass effects through which economic activity would be channelled, providing regionally targeted benefits for the centres and their hinterlands. However, the development of the NSS in Ireland coincided with a shift in emphasis in the planning regime; since the onset of the so-called ‘Celtic Tiger’ era, the Irish planning system had become increasingly development led (Murphy and Fox-Rogers 2015). The neoliberal turn prioritised market-led development, deregulation, public-private partnerships and low corporate taxation (Kitchin et al. 2015). The result was a laissez-faire, uncoordinated approach to housing and planning policy and, in practice, was consistently undermined by localism, clientelism and cronyism (Kitchin et al. 2015).

It is clear that the NSS did not succeed in its objectives. In particular, the spatial patterns of residential and commercial activity have not followed those envisaged in the NSS (Morgenroth 2013). Investment in infrastructure in Ireland lagged behind the rapid rate of development over the decade to 2007. This was particularly evident with new housing provision, whereby housing developments were frequently completed without adequate supporting social and physical infrastructure (McAteer and Stephens 2011). The planning system allowed housing construction to rise from a figure of 19,000 in 1990 to a staggering 93,000 in 2006 (Whelan 2010), for a population of approximately 4.2 million at that time (Central Statistics Office

(CSO) 2007). In comparison, the UK only produced 160,761 for a population of approximately 60 million (for the same year) (Murphy and Fox-Rogers 2015).

The 2006 *National Development Plan (NDP)* recognised that Ireland had a significant infrastructural deficit, which was compounded by unevenly distributed regional development (Grist 2012). However, the NSS was not dovetailed with capital expenditure in the *National Development Plan*, and there was a lack of joined-up planning between local, regional and national strategies (O'Callaghan et al. 2014). While the NSS emphasised the need for future development to be linked to, or concentrated within, primarily the gateways and hubs, the strategy was not given a legislative basis. The NSS was therefore viewed as merely a guidance document for planners and policymakers which proved to be a fundamental weakness (Meredith and van Egeraat 2013). As a strategic framework for the Irish planning system, the NSS was also undermined by inherent weaknesses of the planning system itself. The *2010 Planning and Development (Amendment) Act* only established clear legislative linkages between national policies, regional guidelines, development plans and local area plans (Grist 2012), 8 years after publication of the strategy. Excessive and inappropriately located zonings and developments effectively worked against the implementation of NSS principles and priorities and undermined efficient government investment in infrastructure and services in the period 2002–2007 (Grist 2012). In fact, property development became a dominant driver of economic growth in the years 2000–2007 and a highly significant source of employment (O'Callaghan et al. 2014) as well as government revenue, ensuring a lack of critical discourse on planning and development.

4.2.2.3 Limitations and Shortcomings

Primarily, the NSS was critically undermined by a lack of appropriate governance structures (Breathnach 2010). This was compounded by radical changes in perspective of what the planning system should do with consistent undermining by vested interests (Kitchin et al. 2015). In the period from 2000 to 2007, patterns of physical development diverged significantly from the strategy, as follows (Daly and Kitchin 2013):

- Diffusion of new developments in suburban, peri-urban and rural locations resulting in car-dependent travel patterns
- New housing development frequently outran the provision of essential social and physical infrastructure
- Pro-growth, permissive nature of Irish planning system assisted in inflating the property bubble
- Failure to act as an effective counterbalance to speculative developer-led pressure

The role of government policy and of the planning system in the recent economic crises in Ireland is clear. Irish policy was uncoordinated, piecemeal and favoured development interests, with too few checks and balances, thus enabling a property bubble to rapidly inflate, accompanied by an oversupply of stock (Kitchin et al. 2015). Clearly, planning was not the chief reason for these problems; as described in Fig. 4.1, planning settings contribute to the wider complex of the urban sociotechnical regime, including the wider fiscal context, the availability of cheap credit on international markets and, in the case of Ireland from 2000 on, the adoption of the Euro. However, by enabling unchecked, sprawling development, planning practice in Ireland helped to further exacerbate an overheating housing market, rather than act as an important 'brake' on the market. Urban sprawl has in fact accelerated since the publication of the NSS, and spatial patterns of enterprise activity do not correspond to those planned in the NSS (Morgenroth 2013). Despite an NSS that was lauded as state of the art on publication, and which advocated a principle-based approach to development, development over the period 2002–2007 constituted a free for all in planning terms. As described by Breathnach (2010; p. 1), 'the sprawl of housing and other forms of property development which has peppered the landscape over the last decade would, quite understandably, lead any visitor to the country to conclude that no form of planning of any kind operates in this country'.

4.2.3 The Metropolitan Scale: Strategic Planning in Melbourne, Australia

4.2.3.1 Context for Strategic Spatial Planning

In the Australian context, state governments are key actors in shaping the planning system through the development of strategic plans and goals (every 4 or 5 years) and regulatory frameworks (Planning and Environment Act and land use controls) which guide and direct urban development decisions largely made by local governments. Local governments have limited power to influence the planning regime, but they can develop local planning policies and controls largely in accordance with state government directions. The national level has played a limited role in urban planning beyond developing high level urban policy guidelines and investing in large-scale infrastructure including major road networks.

Emerging out of the late 1980s economic recession, the 1990s in the state of Victoria (and Australia more broadly) were characterised by a shift in public-policy making involving the privatisation of public utilities delivering public transport, energy and water, the deregulation of the planning system and a weakening of the institutional capacity to develop long-term strategic plans for metropolitan and regional areas. One of the factors contributing to political shift away from a conservative agenda in the late 1990s was a groundswell of community concern about unregulated urban development which had been allowed to occur across the metropolitan area with little concern for long-term sustainability. In the early 2000s, a state labor government was elected and immediately started working on a metropolitan plan for Melbourne, which was released in 2002 called *Melbourne 2030: Planning for Sustainable Growth* (DOI 2002). The process for developing the plan involved several years of research, the production of technical reports and rounds of community and stakeholder engagement which informed the development of a vision and objectives for metropolitan Melbourne to guide development for 30 years.

4.2.3.2 Plan Proposals Versus Actual Outcomes

The *Melbourne 2030* plan was applauded by planners for signalling a shift in direction for urban planning away from simply facilitating growth and towards the achievement of social and environmental goals. For example, a number of key goals were included to curtail urban sprawl, first by establishing an urban growth boundary, second by targeting the majority of new housing development within established areas and third by identifying activity centres to concentrate development. *Melbourne 2030* marked a significant shift in vision and purpose for urban planning from the preceding decade of planning which included no long-term sustainability goals for Melbourne.

In 2008, the plan was reviewed and considered to be failing to deliver on most of its objectives (AEG 2008). The clear message was that the plan lacked the institutional and regulatory capacity to implement actions effectively. For example, while the plan stated that a percentage of housing development would be concentrated around major activity centres to minimise sprawl, these targets were not met (Goodman et al. 2010). The urban growth boundary established in 2003 was moved four times over the following decade to allow the outward spread of low-density suburbs with little to no public transport or adequate services. In other words, the dominant model of urban development established in the 1990s continued despite the effort to develop a vision and objectives for a sustainability transition for Melbourne. With no clear targets, measures or resources committed to the achievement of sustainability goals for housing and transport; over time these goals were either ignored or just not implemented. Despite the development of strategic objectives, the urban planning system did not change and has continued to exacerbate sustainability challenges in Melbourne including increasing rates of car usage and congestion, a housing affordability crisis, limited public transport in outer suburbs and a city that continues to expand outward and, in the process, destroy prime agricultural land and biodiversity (Buxton et al. 2016).

Following the failure of *Melbourne 2030* to significantly shift urban development trajectories, there has been mounting criticism of the role and capacity of planning to deliver on sustainability goals in Melbourne, in terms of both the weak regulatory framework and institutional capacity, to deliver on sustainability outcomes. Very little changed following the review of *Melbourne 2030*, particularly as there was a change of state government returning a liberal-national (conservative) coalition to power. A new metropolitan plan, *Plan Melbourne* (DTPLI 2013), was not released until 2014; a document which heralded a return to prioritising economic growth goals with some acknowledgement of the need to consider sustainability outcomes. With a projected population of 9.1 million by 2050 (up from 4.3 million in 2014),

Plan Melbourne is concerned with accommodating that growth within a 'permanent boundary' which means limiting development in some of the fastest growing regions in Victoria and Australia. The Plan is centred on the idea of shifting urban growth from greenfield sites to urban renewal precincts which will be in close proximity to jobs and public transport. While this future-oriented goal is admirable, the focus on 'urban renewal' constitutes an urban growth model that fits the existing regime of facilitating economic development and in reality does little to dissuade the continued growth of low-density suburbs on the urban fringe nor address the problem of housing affordability within existing city boundaries.

4.2.3.3 Limitations and Shortcomings

As a tool to shift the development trajectory towards more low-carbon outcomes, there was little in *Plan Melbourne* to suggest any intent. Some reference is made to the role of small-scale energy precincts (i.e. cogeneration trial projects) and a statement about needing to support local government and the private sector to pursue similar projects. The process for developing the *Plan Melbourne* was quite different to *Melbourne 2030* with limited research and consultation which culminated in the majority of the Ministerial Advisory committee appointed to develop the Plan quitting prior to its release. After years of development, the final plan included a series of objectives, many of which were vague and lacked targets or clear strategies. Since its release there has been a change of government and the Plan is now under review by the current Labor government. Such ad hoc outcomes have also been seen elsewhere in Australia.

4.3 Understanding Planning as Sociotechnical Process

Continued investments in infrastructure with adverse environmental impacts imply that sustainability priorities are not embedded in planning processes (Malekpour et al. 2015). Lawrence et al. (2013) explain this lock-in and highlight that conventional planning approaches are entrenched within current decision-making frameworks; long-term planning follows the path-dependent legacy of those conventional approaches cited in Malekpour et al. (2015). While the language of policy documents and strategic plans has embraced sustainability goals, the legacy of an incremental approach to infrastructure planning, which has been the dominant planning approach for decades, is embedded within planning cultures and is clearly at odds with aspirations for sustainability transitions.

As presented, current policy approaches prioritise development-led economic growth above other sustainability dimensions (social, environmental). In the case of Ireland's spatially polarised economy, this prioritisation does not appear to be afforded a spatial or scaled differentiation; despite political rhetoric, Dublin remains as Ireland's economic engine with the implicit expectation that 'trickle-down' or 'trickle-out' economics will provide an economic lifeline to settlements and regions outside of the capital's sphere of influence. In the cases of both Ireland and Melbourne, state-of-the art policy development approaches, including futures methods, scenario-planning techniques or evidence-based planning, have not been appropriately integrated into the planning process. While it would appear that both the *NSS* and *Melbourne 2030* were prepared with the best available evidence and a progressive remit, the gap between plan development and plan implementation was considerable and fatally undermining in both cases.

It is interesting to note the overt focus on economy, technology and the built environment, with little engagement with people, the dynamics of social mobilisation or community-level capacity to generate change. Collier et al. (2013) argue that it is important to utilise multiple perspectives in the analysis and management of complex systems; local, nonexpert knowledge also has a high value in landscape-level issue management. Such inclusivity is a notable absence in the strategic plans reviewed. Further, a lack of political (and community) buy-in and a prioritisation of economic expediency are common to both cases and demonstrate the current *modus operandi* of the planning system in both Australian and Irish jurisdictions, characterised by short-termism, market-led development and with a tendency to view strategic plans as aspirational, rather than as binding blueprints for development. The lack of political support for the visions articulated in the *NSS* and in *Melbourne 2030* was arguably the single biggest contributor to failure of these strategic plans. This political dimension is important, as is the relationship between power, politics and planning practice. Meadowcroft describes (2011, p. 71) politics as 'the constant companion of socio-technical transitions, serving.…as context, arena, obstacle, enabler, arbiter, and manager of repercussions'. Avelino et al. (2016),[1] Meadowcroft (2011) and Avelino and Wittmayer (2015) all argue for a deeper consideration of politics in transitions process. Here, a treatment of the politics of sustainability transitions in planning systems is beyond the scope of this chapter, but represents a natural progression from our analysis here for future work. For the purposes of this chapter, it is important to note, however, the heteronomy of planners within the current urban sociotechnical regime and the limited agency of planners to enact wider change across the regime.[2]

The agency issue further highlights capacity issues for the urban sociotechnical regime more broadly, currently not configured to deal with the emerging dynamics at the landscape level of the MLP. Global challenges including climate change, energy security and economic stability present new, evolving and risk-laden dynamics to communities across scales, from local and regional to national and international. Thornbush et al. (2013) highlights the need for both the anticipation of external

[1]Sustainability transitions are inherently political, and as encompassing, long-term processes of multiple changes in sociotechnical systems, they require broad understandings of the political (Avelino et al. 2016).

[2]The papers by Ghavami et al. (2016) and Merrie and Olsson (2014) provide deeper insight into the agency of planners for urban land use planning and marine spatial planning contexts, respectively.

shocks as well as an equipped socioeconomic infrastructure to limit future urban vulnerabilities, for instance. Further, while urban planning tools do consider environmental change, natural hazards and phenomena that are influenced by climate change, they often tend to operate within the old planning logic of a conditionally programmed planning approach. Business-as-usual approaches therefore remain predominantly reactive to MLP landscape-level forces. Kirby (2014) argues that political realities have often dictated that proactive measures are not possible, due to vested interests and sunk costs, key determinants of inertia and change dynamics within the regime.

4.4 Transition Management and Steering Planning Transitions

The complexity and uncertainty of a transition makes it difficult, if not impossible, to deliberately engineer (Cohen and Ilieva 2015). The TM literature has been criticised for being undemocratic, unrealistic and over ambitious in nature (Shove and Walker 2007). Yet, solving global environmental crises requires the design and management of sociotechnical transitions, making it important to understand why and how transitions occur (Cohen and Ilieva 2015). The challenge of transition for current planning systems appears to be an organisational and social one, rather than a technological one. Geels and Schot (2007) describe such a transition as a 'Transformation Transition', that is, a change in the sociotechnical regime without recourse to one dominant technology. This is a considerable challenge for planning systems that have worked in incremental mode for so long. At present, the capacity of planning systems and tools to account for differential and dynamically changing vulnerabilities of communities and physical infrastructures exposed to environmental risks remains questionable (Birkmann et al. 2014). Table 4.1 presents an overview of the case study strategic plans, appraised according to the applied Transition Management principles presented in Sect. 4.1.2. Analysis confirm that while some elements of Transition Management are evident in the reviewed plans, there is considerable scope for TM principles to be better integrated into strategic spatial planning, particularly with regard to policy integration, reflexive governance and a wider consideration of stakeholder perspectives.

In addition to the limitations of the planning documents described in Table 4.1, the planning systems in both contexts can be described by the following characteristics: lack of coherence/consistency across government departments and policy documents, lack of principle-based approach in local planning decision-making, incapacity to incorporate multiple perspectives and stakeholder views, short-termism and limited time horizon and outlook for decision-making and little to no consideration of future risk and uncertainty. A more comprehensive and joined-up policy process would enable a wider societal dialogue on the goals of economic development and societal cooperation. Recognition of the landscape level,

Table 4.1 Transition Management features evidence in case study strategic plans

TM principles	NSS	Melbourne 2030	Features of the plans
Long-term thinking	√	√√	Longer-term outlook present [20 years NSS, 30 years Melbourne 2030]. However, limited consideration of risk, uncertainty or future vulnerabilities is evident. The word 'risk' does not appear in the NSS; 'vulnerable' is mentioned twice, in terms of 'vulnerable sections of society' (p. 104) and 'environmentally or visually vulnerable areas' (p. 109)
			The word 'risk' is mentioned twice in Melbourne 2030, in the context of 'investor risk' (p. 72) and water quality risk (p. 101). However. Melbourne 2030 does explicitly state that coastal areas are 'vulnerable to the potential impacts of climate change' (p. 104) and highlights the need for future adaptive strategies. The word 'vulnerable' is used three times in an environmental context and three times in terms of social groups in Melbourne 2030
Integrated policy	**X**	~	Language characterised by terms such as 'integrated planning frameworks' in NSS (p. 118) but little to no details provided on these; Melbourne 2030 mentions 'integrated' over 25 times and does explicitly reference an 'Integrated approach to the achievement of environmental, social and economic outcomes' (p. 167)
			However, disconnect between strategic plans and planning practice evident from recent history in both contexts
Innovation space	~	√	The word 'innovation' appears 24 times in the NSS, with emphasis on the term 'innovation capacity'. However, little detail is provided on how such innovation is to be fostered, other than though balanced regional development
			'Innovation' appears 29 times in Melbourne 2030, with emphasis on the term 'innovation economy' (p. 165). A number of specific measures are detailed under policy 4.4 (p. 167) on how such innovation is to be fostered
Innovation policy support	**X**	√	Innovation not linked to any specific policies in NSS
			Innovation explicitly linked to named policy (policy 4.4.) in Melbourne 2030 (p. 13)
Reflexive governance	**X**	**X**	Words such as 'reflexive', 'governance' and 'feedback' not mentioned in NSS, while Melbourne 2030 does mention 'regular monitoring and review of trends and on feedback as initiatives are implemented' (p. 154) and invites inputs from stakeholders in workshop setting, commitments fall short of reflexive governance practice

(continued)

Table 4.1 (continued)

TM principles	NSS	Melbourne 2030	Features of the plans
Multi-actor approach	~	~	Private sector led development approach and government as facilitator in both cases; however, limited input from civil society
			'Stakeholders' mentioned once in NSS (p. 123) 'relevant stakeholders to work in partnership', but further elaboration on who these are or what the mechanisms for partnership are comprised are not present
			Stakeholders mentioned 12 times in Melbourne 2030, but beyond a workshops approach, mechanisms for engagement are not clear.

Key:
√ = Transition Management principle present and addressed in some capacity (√√ = addressed comprehensively)
X = no evidence for this Transition Management principle in planning document
~ = Transition Management principle somewhat present, but in limited capacity

as articulated in the MLP, would force a policy dialogue on these issues in a more systematic and joined-up manner. Moore et al. (2014) reports the lack of linkage across policies addressing greenhouse gas emission reduction as a large gap. Other examples of policy with scope for better integration include health and well-being targets, transportation and urban design. At present, such debate occurs in a largely ad hoc and undirected manner and without a view on comprehensive response mechanisms. Based on TM principles, a more future-oriented and inclusive debate is required, firstly, on the goals of public policy; secondly, on the concepts, visions and rationale used for the development of this policy; and thirdly on the pathways to achieve outcomes through reflexive governance methods.

4.5 Revaluing Planning: Locating Sustainability Trajectories

In view of the failure of the political and planning systems to deal with the recent economic challenges, with relatively short-term time horizons, the capacity to deal with the large-scale, multi-sector and long-term and chronic problems of climate change, energy security and resilience planning is highly questionable. The recent economic and housing crises in Ireland and Victoria highlight a plethora of deficiencies in governance systems which acted as barriers to a more resilience-focused (economic) planning and development agenda.

To adequately respond to the spatially differentiated impacts of climate change and energy transition, communities will have to develop locally and regionally appropriate approaches. This will require an extensive debate on governance structures and on alternative visions for these; for example, the concept of subsidiarity has not been embraced by Ireland's deeply centralised political establishment and there is at present no meaningful regional layer of governance in the state. In the case of Melbourne, the lack of a metropolitan authority powerful enough to implement strategic plans and coordination of infrastructure provision is a key governance gap. Whatever governance structures are agreed upon, effective decision-making in the strategic sense will require that policy visions and their on-the-ground practical interpretation align for effective implementation, addressing the major disconnect between strategic capacity and operational capacity, currently a key challenge.

While sustainability goals have been debated and applied in planning practice for decades, sustainability has been typically framed as a type of ideal 'end state', a long-term goal to be aimed for. Planning policies to date have largely failed to deliver on these stated goals. The concept of adaptive urban governance underscores that there is a need for a paradigm shift to move from the dominant focus in urban adaptation on the adjustment of physical structures towards the improvement of planning systems, tools and governance processes (e.g. norms, institutional settings, etc.) themselves (Birkmann et al. 2014). Sociotechnical transitions concepts are useful to apply in this context, both in Australia, Ireland and other developed countries as follows: First, current planning approaches perpetuate established sociotechnical configurations by neglecting context uncertainties, by ignoring radical system alternatives and by focusing on narrow value considerations (Truffer et al. 2010). The dynamic nature of the MLP can therefore broaden the scope of planning discourse. Second, the framing of niche-regime-landscape dynamics (niche innovations build up internal momentum, changes at the landscape level create pressures on the regime and destabilisation of the regime creates windows of opportunity for the diffusion of niche innovations (Geels 2014)) presents a valuable theoretical tool when planning practice is faced with rapid technological change and the existential threat of climate change. As demonstrated in Sect. 4.4, TM principles can highlight where current planning practice is deficient and where an expansion of planning's frame of reference can better foster sustainability outcomes.

The need for a coordinating, integrated and joined-up strategic policy has never been more pressing. Ireland's response to climate change is illustrative, having to date been reactive, unambitious and without any semblance of joined-up thinking, across both mitigation and adaption domains. The shifting landscape level also provides substantial opportunities, whereby changing technological and economic conditions provide a means through which to radically alter the problematic elements of the current paradigm. High-speed broadband and the availability of extensive renewable energy resources now provide a fundamental basis for globally connected economic activity in fringe communities, for example, on Ireland's western seaboard.

From a transitions theory perspective, planning as a discipline is inherently space-based. The authors argue that a more in-depth interrogation of sustainability and planning practice could be realised through application of de Haan's (see Chap. 2) *locus* and *nexus* concepts, providing a situated space-based focus through which to further explore processes of uneven and spatially differentiated transitions. This could have valuable practical as well as theoretical utility. Finally, the authors fully recognise the limitations of the framing assumptions of this work; further to Gleeson's description in Chap. 3, the framing for our analysis is most certainly an optimistic one, with a trust that 'the intelligent application of levers of policy' can engender positive change. Whether this assumption is valid or not will make itself evident within a very short time horizon. Current global emission trajectories (UNEP 2016) suggest that Gleeson's (2017) more fatalistic outlook is increasingly becoming the more likely scenario.

4.6 Conclusions

Strategic spatial planning has a significant role to play to guide and direct urban development and land use towards a more sustainable future. Planning processes are vital tools in this context, with the capacity to be anticipatory, forward-looking and coordinating activity across disparate areas of governance, as well as across a wide range of stakeholders with diverse interests and values. We have sought to demonstrate through two cases how planning practice in the case of two market-driven urban sociotechnical regimes is characterised by an overly technical/technocratic bias and also remains largely disconnected from other key dimensions of the urban sociotechnical regime; a key disconnect between planning practice and other policies (fiscal policy, energy policy, etc.) has evidently caused significant problems in achieving sustainability progress. This chapter has forwarded that strategic spatial planning could draw upon sociotechnical transition concepts such as the MLP and TM to expose current shortcomings in the governance of urban sociotechnical regimes and to expand the remit of spatial plans and of planning practice more broadly. Despite this, it is recognised that, in practice, in the context of current political and economic paradigms, this represents a considerable if not impossible challenge. In view of current developmental and economic imperatives and norms, it is difficult to see how planning interventions can truly foster city-region sustainability, when first priority is consistently and uniformly afforded to narrow economic interests. A systematic consideration of strategic policy approaches to address future economy-wide sustainability challenges and the need to urgently foster a transition to a low-carbon future are paradigm shifting challenges which require new means of conceptualising and delivering city regions; such a transition would require transformation in political and economic norms as well as in the planning system and wider urban sociotechnical regime. A move from narrowly defined neoliberal economic orthodoxy represents a first priority. Such a shift could well engender a new environment of innovation, synergy and cooperation, based on place-specific and tailored sustainability principles.

References

Albrechts L (2015) Breaking out of the box: ingredients for a more radical planning. Procedia Soc Behav Sci 184:104–110. doi:10.1016/j.sbspro.2015.05.063

Audit Expert Group (2008) Melbourne 2030 Audit Expert Group Report Melbourne, Department of Planning and Community Development

Avelino F, Wittmayer JM (2015, December) Shifting power relations in sustainability transitions: a multi-actor perspective. J Environ Policy Plan :1–23

Avelino F, Grin J, Pel B, Jhagroe S (2016) The politics of sustainability transitions. J Environ Policy Plan [Internet] 18(5):557–567. Available from: https://www.tandfonline.com/doi/full/10.1080/1523908X.2016.1216782

Birkmann J, Garschagen M, Setiadi N (2014) New challenges for adaptive urban governance in highly dynamic environments: revisiting planning systems and tools for adaptive and strategic planning. Urban Clim 7:115–133. doi:10.1016/j.uclim.2014.01.006

Breathnach P (2010) The National Spatial Strategy update – more of the same old "same old." In: Irelandafternama.wordpress.com.https://irelandafternama.wordpress.com/2010/10/14/the-national-spatial-strategy-update-%E2%80%93-more-of-the-same-old-%E2%80%9Csame-old%E2%80%9D/. Accessed 6 June 2015

Bridge G, Bouzarovski S, Bradshaw M, Eyre N (2013) Geographies of energy transition: space, place and the low-carbon economy. Energy Policy 53:331–340. doi:10.1016/j.enpol.2012.10.066

Buxton M, Goodman R, Moloney S (2016) Planning Melbourne: lessons for a sustainable city. CSIRO Publishing, Melbourne

Central Statistics Office (CSO) (2007) Principal Demographic Results 2006. In: Census 2006. http://www.cso.ie/en/media/csoie/census/documents/Amended,Final,Principal,Demographic,Results,2006.pdf. Accessed 2 Feb 2016

Coenen L, Benneworth P, Truffer B (2012) Toward a spatial perspective on sustainability transitions. Res Policy 41(6):968–979

Cohen N, Ilieva RT (2015) Transitioning the food system: a strategic practice management approach for cities. Environ Innov Soc Trans:1–19. doi:10.1016/j.eist.2015.01.003

Colin Buchanan & Partners (1968) Regional studies in Ireland. An Foras Forbatha, Dublin

Collier MJ, Nedovic Z, Aerts J et al (2013) Transitioning to resilience and sustainability in urban communities. Cities 32:S21–S28. doi:10.1016/j.cities.2013.03.010

Daly G, Kitchin R (2013) Shrink smarter? Planning for spatial selectivity in population growth. Administration 60:159–186

Department of Infrastructure (DOI) (2002) Melbourne 2030: planning for sustainable growth, Victorian Government, Melbourne

Department of Transport Planning and Local Infrastructure (DTPLI) 2013 Plan Melbourne: metropolitan planning strategy, Victorian Government, Melbourne

Eraydm A (2012) Contradictions in the neoliberal policy instruments: what is the stance of the state? In: Tasan-Kok T, Baeten G (eds) Contradictions of neoliberal planning, GeoJournal. Springer, Dordrecht

Gazheli A, Antal M, van den Bergh J (2015) The behavioral basis of policies fostering long-run transitions: stakeholders, limited rationality and social context. Futures 69:14–30. doi:10.1016/j.futures.2015.03.008

Geels FW (2012) A socio-technical analysis of low-carbon transitions: introducing the multi-level perspective into transport studies. J Transp Geogr 24:471–482

Geels FW (2014) Regime resistance against Low-Carbon Transitions: introducing politics and power into the multi-level perspective. Theory, Cult Soc [Internet] (May 2013):0263276414531627. Available from: http://tcs.sagepub.com/content/early/2014/06/27/0263276414531627.full

Geels FW, Schot J (2007) Typology of sociotechnical transition pathways. Res Policy 36:399–417

Ghavami SM, Taleai M, Arentze T (2016) Socially rational agents in spatial land use planning: A heuristic proposal based negotiation mechanism. Comput Environ Urban Syst [Internet]. Elsevier Ltd 60:67–78. Available from: http://dx.doi.org/10.1016/j.compenvurbsys.2016.08.004

Gleeson B (2017) A dangerous transition to hope. In: Moore T, de Haan F, Horne R, Gleeson B (eds) Urban sustainability transitions: Australian cases – international perspectives. Springer, Sydney

Gleeson B, Low N (2000) Revaluing planning rolling back neo-liberalism in Australia. Prog Plann 53(2):83–164

Goncalves J, Antunes Ferreira J (2015) The planning of strategy: a contribution to the improvement of spatial planning. Land Use Policy 45:86–94. doi:10.1016/j.landusepol.2015.01.020

Goodman R, Buxton M, Chhetri P, Taylor E, Wood G (2010) Planning and the characteristics of housing supply in Melbourne, AHURI Final Report No. 157, Australian Housing and Urban Research Institute, RMIT Research Centre

Grist B (2012) The Irish National Spatial Strategy. In: Planning for states and nation/states: a TransAtlantic exploration. University College Dublin, Dublin, pp 1–12

Gunder M (2006) Sustainability: planning's saving grace or road to perdition? J Plan Educ Res 26:208–221. doi:10.1177/0739456X06289359

Hansen T, Coenen L (2015) The geography of sustainability transitions: review, synthesis and reflections on an emergent research field. Environ Innov Soc Trans 17:92–109. doi:10.1016/j.eist.2014.11.001

Hughes B (2014) Demographic growth in Ireland since 2011: some geographic implications for future spatial planning, housing and infrastructure provision, Dublin

Hughes B, Sirr L (2015) Why balanced regional development should be replaced by urban agglomeration initiatives in Ireland's reconfigured National spatial and economic strategy, Dublin

Jabareen Y (2015) Geoforum city planning deficiencies & climate change – the situation in developed and developing cities. Geoforum 63:40–43. doi:10.1016/j.geoforum.2015.05.017

Kemp R, Loorbach D (2006) Transition management: a reflexive governance approach. Reflexive governance for sustainable development. Edward Elgar, Northampton, pp 103–130

Kemp R, Rotmans J (2009) Transitioning policy: co-production of a new strategic framework for energy innovation policy in the Netherlands. Policy Sci 42:303–322

Kern F, Rogge KS (2016) The pace of governed energy transitions: agency, international dynamics and the global Paris agreement accelerating decarbonisation processes? Energy Res Soc Sci 22:13–17. doi:10.1016/j.erss.2016.08.016

Kirby A (2014) Current research on cities. Cities 41:S1–S2. doi:10.1016/j.cities.2014.06.009

Kitchin R, Hearne R, O'Callaghan C (2015) Housing in Ireland: from crisis to crisis, Maynooth

Kumar P, Geneletti D (2015) How are climate change concerns addressed by spatial plans? An evaluation framework, and an application to Indian cities. Land Use Policy 42:210–226

Lawrence J, Reisinger A, Mullan B, Jackson B (2013) Exploring climate change uncertainties to support adaptive management of changing flood-risk. Environ Sci Pol 33:133–142

Leshinsky R, Legacy C (eds) (2015) Instruments of planning: tensions and challenges for more equitable and sustainable cities. Routledge, New York

Malekpour S, Brown RR, De Haan FJ (2015) Strategic planning of urban infrastructure for environmental sustainability: understanding the past to intervene for the future. Cities 46:67–75. doi:10.1016/j.cities.2015.05.003

McAteer B, Stephens S (2011) Town centre management: a solution to the challenges facing urban centres in Ireland? J Place Manag Dev 4:264–271. doi:10.1108/17538331111176075

Meadowcroft J (2011) Engaging with the politics of sustainability transitions. Environ Innov Soc Trans 1:70–75. doi:10.1016/j.eist.2011.02.003

Meredith D, van Egeraat C (2013) Revisiting the National Spatial Strategy ten years on. Administration 60:3–9

Merrie A, Olsson P (2014) An innovation and agency perspective on the emergence and spread of Marine Spatial Planning. Mar Policy [Internet]. Elsevier 44:366–374. Available from: http://dx.doi.org/10.1016/j.marpol.2013.10.006

Moore T, Horne R, Morrissey J (2014) Zero emission housing: policy development in Australia and comparisons with the EU, UK, USA and California. Environ Innov Soc Trans 11:25–45. doi:10.1016/j.eist.2013.12.003

Morgenroth E (2013) Economics – the missing link in the National Spatial Strategy. Administration 60:41–59

Murphy JT (2015) Human geography and socio-technical transition studies: promising intersections. Environ Innov Soc Trans 17:73–91

Murphy E, Fox-Rogers L (2015) Perceptions of the common good in planning. Cities 42:231–241. doi:10.1016/j.cities.2014.07.008

Normann HE (2017) Policy networks in energy transitions: the cases of carbon capture and storage and offshore wind in Norway. Technol Forecast Soc Change. doi:10.1016/j.techfore.2017.02.004

NSS (2002) National Spatial Strategy for Ireland: people, places and potential. Dublin

O'Callaghan C, Boyle M, Kitchin R (2014) Post-politics, crisis, and Ireland's "ghost estates". Polit Geogr 42:121–133. doi:10.1016/j.polgeo.2014.07.006

Raven R, Schot J, Berkhout F (2012) Environmental innovation and societal transitions space and scale in socio-technical transitions. Environ Innov Soc Trans 4:63–78. doi:10.1016/j.eist.2012.08.001

Rotmans J, Loorbach D (2008) Transition management: reflexive governance of societal complexity through searching, learning and experimenting. Managing the transition to renewable energy, 15–46

Rydin Y (2013) The future of planning. Policy Press, Bristol

Sager T (2011) Neo-liberal urban planning policies: a literature survey 1990-2010. Prog Plann [Internet]. Elsevier Ltd 76(4):147–199. Available from: http://dx.doi.org/10.1016/j.progress.2011.09.001

Shove E, Walker G (2007) CAUTION! Transitions ahead: politics, practice, and sustainable transition management. Environ Plan A 39:763–770. doi:10.1068/a39310

Thornbush M, Golubchikov O, Bouzarovski S (2013) Sustainable cities targeted by combined mitigation – adaptation efforts for. Sustain Cities Soc 9:1–9. doi:10.1016/j.scs.2013.01.003

Truffer B, Störmer E, Maurer M, Ruef A (2010) Local strategic planning processes and sustainability transitions in infrastructure sectors. Environ Policy Gov 20(4):258–269

UNEP (2016) The emissions gap Report 2016. United Nations Environment Programme. doi:ISBN 978-92-807-3617-5

Walsh JA (2013) The National Spatial Strategy: rationale, process, performance and prospects. Administration 60:13–40

Walsh C (2014) Rethinking the spatiality of spatial planning: methodological territorialism and metageographies. Eur Plan Stud 22:306–322. doi:10.1080/09654313.2012.741568

Whelan K (2010) Policy lessons from Ireland's latest depression. Econ Soc Rev (Irel) 41:225–254

Wilson E, Piper J (2010) Spatial planning and climate change. Routledge, New York

Chapter 5
How Could Sustainability Transition Theories Support Practice-Based Strategic Planning?

Judy Bush, Lu Aye, Dominique Hes, and Paul Murfitt

Abstract Theories of sustainability transitions aim to explain the processes, pathways and actors that are involved in transformations in technologies and practices. Whilst there is a growing body of research developing theoretical understandings, there has been less documented on how theories are utilised and applied by practitioners themselves.

This chapter reports on a case study that investigated whether provision of targeted information on theories of sustainability transitions could strengthen organisational strategic planning. If planning is informed by transition theories, would this assist and strengthen organisational visioning, ambition and confidence? The research focuses on Moreland Energy Foundation Limited (MEFL), a community-based not-for-profit organisation working on sustainable energy and climate change action in Melbourne, Australia. During 2014–2015, MEFL developed a new strategic plan. As part of this process, theories of sustainability transitions were presented to the organisation's Board and staff, to support the strategic planning and to investigate the theories' roles in the planning process.

It was found that inclusion of the sustainability transitions theoretical framework led to the organisation explicitly defining its shared 'model of change', reinforcing the organisation's conceptualisation of its role as an 'intermediary' between grassroots and governments. The process demonstrated the potential impact of

J. Bush (✉)
Department of Infrastructure Engineering and Thrive Research Hub, The University of Melbourne, Parkville, Australia
e-mail: judy.bush@unimelb.edu.au

L. Aye
Renewable Energy and Energy Efficiency Group, Department of Infrastructure Engineering, The University of Melbourne, Parkville, Australia
e-mail: lua@unimelb.edu.au

D. Hes
Thrive Research Hub, The University of Melbourne, Parkville, Australia
e-mail: dhes@unimelb.edu.au

P. Murfitt
Moreland Energy Foundation Ltd 2008–2016, Brunswick, Australia
e-mail: murfitt.paul@gmail.com

T. Moore et al. (eds.), *Urban Sustainability Transitions*, Theory and Practice of Urban Sustainability Transitions, DOI 10.1007/978-981-10-4792-3_5

research-practice partnerships in strategic planning. However the findings also highlighted the continuing challenges of connecting research and practice.

Keywords Research-practice partnerships • Multilevel perspective • Climate change action • Strategic planning • Non-government organisation • Intermediaries

5.1 Introduction

Theories of sustainability transitions aim to explain the processes, pathways and actors that are involved in the transformations in technologies and practices for sustainability. There is a growing body of research focused on the role of community organisations and grassroots groups in initiating and implementing climate change planning and actions towards sustainability transitions (e.g. Kent 2016; Seyfang and Haxeltine 2012). To date, however, only a limited number of examples have been documented, mostly in European countries, of how theories can be applied by practitioners themselves (Narberhaus 2011; Nevens et al. 2013).

This chapter reports a case study that investigated the extent to which transition theories can be translated into practical information to be utilised by a community organisation, Moreland Energy Foundation Limited (MEFL), in its work on sustainability transitions for climate change action. If organisational planning is informed by transition theories, would this assist and strengthen organisational visioning, ambition and confidence?

We propose that the potential benefits of incorporating sustainability transition theories into strategic planning could include reinforcing a co-focus on long-term vision and short-term implementation (Rotmans et al. 2001); providing a basis for locating the organisation within a global community of researchers and practitioners (Bulkeley et al. 2015); encouraging and legitimising the organisation's work on (currently) non-commercial niche innovations (Castán Broto and Bulkeley 2013; Schot and Geels 2008); highlighting the importance of stakeholder analysis, mapping, partnerships and alliances (Loorbach 2010); encouraging continuing climate change mitigation efforts in the face of inadequate action at higher levels of government; and strengthening links between research and practitioners (Nevens et al. 2013) and between theory and practice (Loorbach 2014).

This chapter outlines the context in which the research was undertaken, including an overview of MEFL's work, within the broader setting of climate change planning and action in Australia. It describes the research methods and highlights the key findings, with discussion of how research and practice can improve the linkages between each group, and the effectiveness of collaborative efforts. It concludes with reflections on how this study contributes to urban transitions research and practical action.

5.2 Background

Within Australia, national policy on climate change has become politically polarised, with policies and government approaches shifting at each federal election since at least 2005 (Eckersley 2015). However throughout this period there have been significant achievements in climate change planning and action at the municipal (local government authority) level (Moloney and Horne 2015). These include setting ambitious emission reduction targets and reducing energy use of municipal buildings through energy efficiency retrofits and renewable energy installations (Moloney and Horne 2015). Local action on climate change has been the subject of a number of studies. To date, research has largely focused on the governance and partnerships between state and local governments, and community organisations (McGuirk et al. 2015; Moloney and Horne 2015), as well as innovative governance approaches such as 'regional greenhouse alliances' (Moloney and Fünfgeld 2015). Local- and city-scale climate change action provides a fruitful context for research on how multilevel governance perspectives contribute to understandings of urban transition processes (Bulkeley and Betsill 2013), as well as adding to a growing picture of how city-scale action is contributing to tangible action and growing ambition (Wiseman et al. 2014).

It is within this context that we sought to explore how theories of sustainability transitions could be utilised in practice, by practitioners. Transition Management scholars have suggested that the theories, in addition to their work of analysing existing or past transitions, may contribute to empowering or supporting current and future transitions, though there is ongoing debate about how successful or effective this can be (Avelino et al. 2016; Nevens et al. 2013; Shove and Walker 2007). The study aimed to investigate both how the theories can be translated to practice and how they can assist and be applied by practice.

Theories of sustainability transitions have developed rapidly during the last 15–20 years to study the dynamics of sustainability transitions and to identify the barriers and challenges in the adoption of sustainability innovations. The causes for lack of progress in climate action and the active resistance to renewable energy and energy efficiency adoption and uptake have been investigated by both researchers (e.g. Geels 2014) and practitioners (e.g. ACF 2016; Ward and Power 2015). However, there has been little 'crossover' between the two domains and a lack of integration of thinking between the two bodies of analysis. The work of the Dutch Research Institute for Transitions (DRIFT) in a European context provides some of the few existing examples (e.g. Frantzeskaki et al. 2014; Nevens and Roorda 2014; Wittmayer et al. 2014). In addition, whilst research has focused on how transition theories can analyse progress on problems, there has been relatively little research focused on the change and transitions of, and within, the organisations working on these problems or how the theories can support or facilitate their work.

The complexity, severity and urgency of climate change impacts and broader sustainability issues and challenges loom large in the minds of researchers, environmental organisations and community groups (Hamilton 2010; Harre 2011; Hes and Du Plessis 2015). Community-based organisations working on climate change action have acknowledged the struggle to maintain enthusiasm, motivation and momentum amongst their members and the broader community in the face of inaction or active resistance to climate action at higher levels of government, business and media. For example, the Climate Reality project, in partnership with the Australian Conservation Foundation, one of Australia's largest environmental non-government organisations, acknowledged the potential psychological impacts of dealing with climate change and published tip sheets on 'Coping with climate change distress' and 'Dealing with burnout' (CRPA 2016). Other organisations also offering support to climate change activists and organisations include Psychology for a Safe Climate (PSC 2016) and the Australian Psychological Society (APS 2016). Sustainability transition theories may potentially provide an additional means to support these organisations' work, through linking their efforts to others' research and practice-based work from across the globe, as well as to identify the common elements that are preventing or hampering progress and uptake of innovation.

We collaboratively explored the application of transition theories in practice, by practice, in partnership with MEFL in Melbourne Australia. (MEFL's CEO at the time the research was undertaken is a co-author of this chapter, having contributed to furthering the research analysis and outcomes but was not involved in data collection to minimise potential influence.) MEFL is a community-based not-for-profit organisation working towards 'purposive transitions' in sustainable energy use and generation, to contribute to broader climate change mitigation efforts. This is expressed in its organisational vision of 'an active, inspired community tackling climate change with sustainable energy solutions' (MEFL 2015, p. 4). MEFL was formed in 2000 by Moreland City Council, a local government authority located in the northern metropolitan region of Melbourne. MEFL's objective is to reduce greenhouse gas emissions in the Moreland municipality, by assisting its local community to adopt sustainable energy solutions (MEFL 2015). It aims to harness local communities' and businesses' concern to see tangible climate change action and translate this into significant emission reductions at the local and regional level, to influence, create and lead the way for targeted sustainability transitions. In doing so, it seeks to demonstrate climate change leadership to influence other communities across Australia more broadly and to trial new approaches that can be scaled up to broader approaches.

MEFL receives funding from Moreland City Council, supplemented by grants funding from state and federal governments, and fee-for-service consultancies, the latter demonstrating that the experience that MEFL can bring to others is valued. In 2012, MEFL also established a social enterprise, Positive Charge, 'with a mission and purpose to deliver carbon emission reductions at scale across a number of local government areas', expanding its business models and scale of delivery and diversifying its income sources (Positive Charge 2016). MEFL sees itself as a leading organisation in developing and implementing community-based actions (MEFL 2015 p. 6). It points to its relatively long experience in this space and its almost unique model of delivery. It has strong connections both with grassroots organisations and local community groups, as well as with local and higher levels of government bureaucrats and politicians. As such, it serves as a linking organisation or 'intermediary' between sectors and between government and civil society. It has largely been able to maintain a positive, energised focus in the face of government policy and funding uncertainty and instability. Nonetheless, MEFL's staff members (and those in other community-based environmental non-government organisations) acknowledge the challenges of working on climate change action in this sometimes 'toxic' policy and government context, in which attitudes towards climate change responses are polarised and policy and programme approaches lack stability and certainty.

5.3 Methods

This research employed a single case study design (Yin 2014) to investigate how sustainability transition theories could inform and support MEFL's organisational strategic planning. The research utilised the occasion of a strategic planning workshop to opportunistically test the research questions. The opportunity arose through the part-time employment by MEFL of one of the researchers. It was made possible through MEFL's openness and willingness to participate in a research project to explore links with academic research processes and to experiment with new inputs and ideas into its strategic planning process.

During 2014–2015, MEFL developed its 2015–2018 strategic plan. The development of the plan involved a process spanning 18 months and included discussions at Board and staff meetings, consultation with MEFL's partners and stakeholders, preparation of discussion papers and review of the implementation of the previous strategic plan. This preparatory work culminated in a strategic planning workshop, attended by 13 members of MEFL's Board and key staff, held in December 2014. The purpose of the workshop was to formally reach agreement on the strategic priorities and to plan associated actions. Prior to the workshop, the Board and its Governance Committee had already made substantial progress on identifying strategic principles and priorities, and negotiating treatment of issues that had been organisationally 'contentious', or on which there were a range of differing views. It had already been agreed before the December 2014 strategic planning workshop that the organisation's vision, mission and organisational values would remain unchanged.

In preparation for the workshop, a 'background reading pack' was circulated to workshop participants. The reading pack included a four page document, prepared by this chapter's authors, summarising key elements of sustainability transition theories (Markard et al. 2012). The theories that were included were the 'multilevel perspective' (Geels 2011), to locate MEFL's work and role within a broader multilevel context; Transition Management (Loorbach 2010; Rotmans et al. 2001), to highlight procedural and tactical elements; and pathways of sustainability transitions (Geels 2002), to reinforce the role of niche innovation, experimentation and collaboration with other innovators. As well as an explanation of the theoretical material, the reading pack included a short section outlining how the theories could potentially support MEFL's strategic planning process. It also proposed to review and report on the process as a case study research project.

The workshop opened with a 1 h presentation by one of the authors and discussion on the theories of sustainability transitions and practical examples of how they could be applied to MEFL's context and MEFL's work. The inclusion of this session in the workshop agenda was framed as theoretical information that could support MEFL's strategic and business planning. The inclusion of the theoretical material as part of the strategic planning workshop was approached with care and some hesitation by those responsible for organising the workshop. There was concern that there would be resistance or even antagonism towards theoretical, 'academic' material. Therefore, the background paper and workshop presentation were carefully crafted to reinforce the practical aspects of the theories, and the ways that they could be applied to practice in general and to MEFL's context specifically.

Following the workshop, workshop participants were invited to voluntarily participate in semi-structured interviews. The interviews focused on participants' understandings and perceptions of transition theories and how these interacted with strategic planning discussions. Ten members of MEFL's staff and Board, who had all attended the strategic planning workshop, were interviewed. The interviews were held within 8 weeks of the strategic planning workshop but before the strategic plan was finalised and launched. The interviews were recorded and transcribed. Confidentiality of participants was protected, with each participant's transcript identified only with a unique number (e.g. Respondent 3).

Analysis of the interview transcripts used grounded theory techniques to identify the key categories and themes. Grounded theory was used to allow emergent themes from the interview participants to arise through the analysis process (Charmaz and Belgrave 2012). This approach reinforced the importance of the interview participants' voices and reflections to explore and respond to the research question of how theories could support organisational strategic planning, rather than using the researchers' own views on how theories could support practice. The following section reports the results of this analysis; subheadings identify the key themes. Direct quotations from interview participants are included to highlight key themes. The results of the research were reported back to the participants in the form of an earlier draft of this chapter; participants were provided the opportunity to respond or comment, but none were received.

5.4 Results

5.4.1 Familiarity and Understanding of Theories

All respondents said that the theoretical information had been presented in such a way to make it easily understandable and relevant to MEFL's context. Most interview participants reflected that they found the sustainability transition theories relevant and even familiar. The theories were 'very familiar to me, more in a concept of innovation' (Respondent 2). Respondents reflected that the theoretical material drew out and made explicit that which had been assumed and implicit. 'I guess the broad concepts weren't unfamiliar. They were sort of intuitive, but I hadn't thought about them that explicitly' (Respondent 4). The theories 'kind of reflects a mental model that MEFL works with anyway, so it just makes overt what sat in people's thinking' (Respondent 8).

Both the background paper and the workshop presentation reproduced several diagrams to illustrate the theories, including the 'multilevel perspective' as a nested hierarchy (Geels 2002 p. 1261), dynamic transition pathways (Geels 2002 p. 1263) and the Transition Management cycle (Loorbach 2010 p. 173). A number of respondents commented that the diagrams were helpful in their understanding of the theories: 'The thing I found most useful in the info was the picture, with the top bit, the middle bit and the bottom bit ... and the niche stuff breaking through, and particularly having some failed innovations was nice, because we all know failure is part of it' (Respondent 6), and Respondent 9 said of the material presented, 'I love a diagram, I think very visually, I like it because it communicates an approach. For me when we talk about a theory it's actually the diagrams ... I probably should have read the thing'. However, not all participants found the diagrams helpful: one respondent found the material 'like something out of a physics paper' and added 'draw another model, because it's the simple ones that really fit' (Respondent 8).

5.4.2 Applicability of Theories to Strategic Planning Processes

The theoretical presentation, coming at the start of the strategic planning workshop, was seen by participants to provide the framing within which the day's discussions would proceed. Respondent 2 said 'it was a nice reflective, context-setting theoretical start to the day. It's always good to have something a little different to start the day, something a little different, a little abstract to take you out of the day-to-day, whatever that might be. There are a lot of ways you can do that. It was a nice contribution to get people into that space. We're not in the office, we've got to think a little differently today, a theoretical piece encourages reflection, bigger picture thinking, open thinking'. Respondent 3 said 'A nod to the history and a nod to the future, it makes it real, to avoid being vague and abstract, to make the link between theory and practice'. These comments reflected that provision of research theories was an alternative framing for organisational strategic planning (not the 'usual' approach to such workshops: 'I think it helps you stop and think about it, from a different perspective' (Respondent 9)), as well as the effectiveness of the theories in taking participants' focus above or beyond day-to-day concerns, enabling or encouraging a 'strategic' focus for discussions.

5.4.3 Applicability of Theories to MEFL's Work

There was a range of different or divergent views regarding what the theories 'were for' or how the theories applied to MEFL's work. Whilst many suggested that the theories contributed a shared language about change and transition processes for the strategic planning process, participants 'connected' with different aspects of the theoretical material that had been presented. 'It located MEFL's role as niche innovator and how we fitted into bigger picture. It didn't enhance or limit ambition, just located it, placed MEFL's role, how we fitted into bigger picture' (Respondent 3). Some stated that the theories particularly highlighted the importance of partnerships and collaboration. Others saw value in how the theories could support project planning or monitoring and evaluation. Several highlighted the significance of innovation thinking: 'it reinforces that innovation stuff, you have to be innovative to make change so you need to be a bit bold and brave and courageous to make change. That there will be failures in that, and that's ok, and that some of them will break through and make real change' (Respondent 6). 'We may call it innovative approaches to sustainable energy use, but what we're essentially trying to do is sustainable transitions' (Respondent 1). Some questioned the theories' applicability to practice, in particular whether the theories could practically support, inform or direct strategic planning. 'Can it of itself drive a strategy, I'm not sure that it does that' (Respondent 2). Respondent 9 said 'I'm very much a practitioner, how does it get applied, otherwise it's interest ... it's just an academic exercise, whereas if there's something coming out of it then how does it become a tool for the planning phase and how is it influencing that'. These comments highlighted the view that several participants held that for the theories to be used in practice, the development of additional specific, practical tools and a demonstration of their application were needed.

5.4.4 Research Contributions to Practice

Many of the respondents said that academic theories, and research partnerships with universities, were important to reinforce and provide legitimacy to MEFL's approach to climate change action. 'On the one hand it was almost validation for what MEFL does. That was the first instinct that came to me, was, that's what we do' (Respondent 1). University involvement added 'credibility' to MEFL's strategic planning. 'Collaborating with universities gives you credibility, broader recognition' (Respondent 6). Another suggested 'The relationship with the research community, there's a two way benefit: rigour, from researchers to practice, and relevance, from practice to researchers. You need both, cross-fertilisation' (Respondent 3).

In terms of research-practice partnerships, whilst many respondents reflected on the 'credibility' that involvement in university research can bring to the organisation, there was also a sense of distance between the languages, cultures, objectives and time frames of the research community and practice community. 'Bridging between theory and practice is a specialised and difficult task' (Respondent 3). One respondent expressed active distrust of some research processes, particularly based on previous experience as an environmental activist involved in direct action protests, where a student project had sought to intrude on the complex, multilayered inter-organisational planning and implementation of a blockade action; 'the catch is, is it a burden without the outcome' (Respondent 9). Another suggested that 'A lot of activists shy away from any consideration of a theoretical framework' (Respondent 3). One respondent admitted to feeling concern about the inclusion of theories in a practically focused organisation's discussions: 'my overriding focus was to make it non-academic-y' (Respondent 7). The main means for achieving this was to 'turn them into practical plain language structures' (Respondent 7).

5.4.5 Theories' Influence on MEFL's Strategic Plan

Participants agreed that the inclusion of theories provided a different framing for the start of the strategic planning workshop, but their impact on organisational thinking and discussion was limited by not explicitly returning to the material throughout the workshop. Respondent 2 said 'it helps with the bigger stuff, but in terms of setting priorities I'm not sure how it helps with that'. Beyond the workshop, there was agreement to define MEFL's 'model of change' within the strategic plan. 'Even just saying, 'oh ok we need to have our model of change in our strategic plan' wouldn't have happened unless we'd had that discussion at the outset. And I think it's a really good discussion for the organisation to have, to be explicit about it. You put it down and that's when people then say, 'oh no that's not quite right' and that's great, that's when good discussions start to happen' (Respondent 6). Within the finalised strategic plan, MEFL defined its location within the 'multilevel perspective' (Fig. 5.1) and its role as seeking to influence the space between regime and niche (MEFL 2015 p. 11). As such, MEFL adapted and adjusted the nested hierarchy of the 'multilevel perspective' to reflect its own understanding of its location and role in the nested system, thus internalising the theoretical framework.

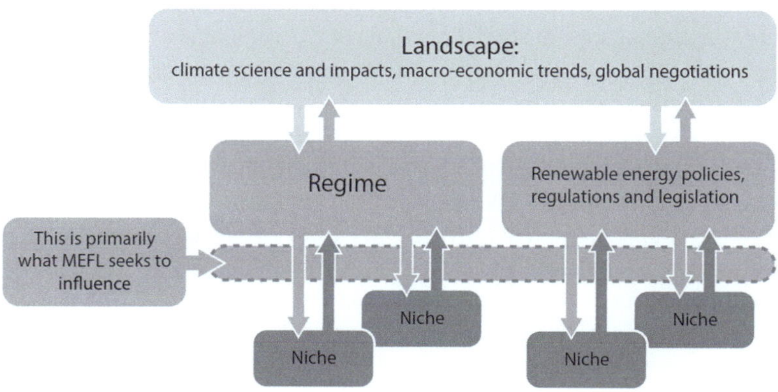

Fig. 5.1 MEFL's representation of its 'intermediary' location within the multilevel perspective (Source, MEFL 2015, p. 10)

It is interesting to note that MEFL defined its role as an 'intermediary', between regimes and niches, even though research on intermediaries had not been presented as part of the theoretical background material.

5.5 Discussion

5.5.1 *Critique of Outcomes: Implications for Theory and for Practice*

As a direct result of inclusion of the theories in planning discussions, MEFL's 'model of change', previously implicit and assumed, was made explicit in its strategic plan (MEFL 2015). MEFL's Board had first considered developing a 'model of change' at a prior strategic planning day (November 2012). However, at subsequent planning days, this discussion had been 'superseded' by more specific and immediate debate focusing on clarifying emission reduction targets and analysis of sectoral opportunities. Discussions at subsequent strategic planning workshops were framed at a more practical, tools-based level, focused on particular aspects of MEFL's operations, including stakeholder engagement, business models and emissions reduction target. The presentation of theories of sustainability transitions supported a shift to a more global, strategic level of discussion and understanding. MEFL's 'model of change' was located within a theoretical framework in which the language and concepts that were directly drawn from sustainability transitions theories, particularly the multilevel perspective, were explicitly defined. This strengthened the organisation's own framing of its role as a connector between niche-level community organisations and households and regime-level state and local governments and businesses. 'The theories provided a strategic framework. The plan is more internally coherent' (Respondent 7).

However, because the theories were not introduced until just before the strategic planning workshop, and substantial previous work had already been undertaken by the Board and staff on refining strategic principles, identifying priorities and scoping associated targets for the plan, the theories did not affect much of the plan's detail. Respondent 7 suggested that 'the theories probably didn't change the high level strategic priorities, but perhaps they impacted the middle level – how the strategic priorities are implemented'. The limited time frame in which the theories were first introduced and discussed restricted the level of engagement of participants with the concepts, languages and applications of the theoretical information. 'The theoretical input was perhaps too limited. On the day, there was the presentation on the theories, and then an attempt to locate MEFL's practice within the theories, but then the theories were not formally returned to in the agenda or discussions throughout the rest of the day' (Respondent 7). The theoretical input was to some extent 'quarantined' to discrete sections within the workshop agenda, in part due to concern about how participants would respond to the 'academic' material. It may be expected that if the theories had been brought into the strategic planning process earlier, they may have had greater impact. Earlier integration of theoretical material would require already well-established research-practice partnerships.

Given the relatively late stage at which the theoretical material was introduced to strategic planning participants, its impact is likely to have been smaller than if it was interwoven into the process from a much earlier stage. The theoretical input contributed to constructing a shared language, which was used during the workshop, and in the final text of the strategic plan. However, the theories' influence to the organisation may have been relatively short-lived, unless the language and concepts are actively and frequently used by organisational members in subsequent implementation of strategic plan priorities and in monitoring and evaluation of implementation. Unless the concepts and language are internalised by participants, there may be little evidence of their significance to the organisation in the future, beyond a 'model of change' reference in a written document. For impacts to extend beyond the end of specific processes, ongoing reinforcement and ongoing relationships and collaborative opportunities between research and practice are required.

Sustainability transition theories could potentially inform and direct ongoing review, monitoring and evaluation of both the implementation of the strategic plan and of broader organisational processes and achievements. However, for this function to be 'operationalised', significant additional input, ongoing collaboration and development of specific tools and guidance material are required.

5.5.2 Research-Practice Collaborations

Loorbach (2014 p. 68) called for new types of research methods for the emerging field of transitions research 'that have an integrative nature, are normative in their ambitions, have a desire to contribute to societal change, and are participatory'. This case study makes a contribution to this field, by seeking to address these methodological elements. The research was undertaken in a form that could be described as a short-term action research project, where action research is 'the collaborative production of scientifically and socially relevant knowledge, transformative action and new social relations' (Wittmayer et al. 2014 p. 468). Whilst the planning and implementation of the research was undertaken collaboratively between researchers and practitioners, the period of direct interaction between researchers and the organisation's Board and staff spanned no more than 4 months, from the time the background paper was circulated in November 2014, through the workshop, subsequent interviews of workshop participants and drafting of the plan, to the Board's endorsement of the finalised strategic plan in March 2015.

Nonetheless, the process demonstrated the potential impact of research-practice partnerships in strategic planning for sustainability transitions and climate change action. However the findings also highlighted the continuing challenges of bridging, or connecting research and practice, and the highly specialised communication and partnership skills required to strengthen these connections. The barriers inhibiting information exchange and partnerships between research and practice are reinforced by the two sectors' different languages, objectives and drivers. 'The hard part is … the barriers are that there are two different agendas, or two different operating systems that people have to function with' (Respondent 8). As Respondent 9 commented, 'it's an exercise in trying to understand each other's objectives. With all good intentions we didn't get anywhere with [a previous research partnership proposal]'.

Actively undertaking and embracing partnership approaches requires explicitly building and maintaining respectful and productive relationships, through which all participants expect to learn and develop (Frantzeskaki and Kabisch 2016). Past experiences can powerfully shape future attitudes and approaches, and developing respectful collaborative relationships requires ongoing contact and communication. Communicating the results of research to the organisation in which the research was undertaken is an important element of developing and maintaining these relationships. The converse, where results are not communicated (even in the case of student projects) to organisations, contributes to erosion of trust and hesitancy in entering into further research projects.

Research-practice collaborations need to provide value or benefit to both parties, preferably through bringing skills and experience that fill gaps in each other's expertise and existing resources (Fig. 5.2). Binder et al. (2015) highlighted key elements that facilitate successful partnership efforts, including establishing and maintaining 'co-leadership' that encompasses clearly defined roles and responsibilities, alignment of goals and a good communication structure.

In addition, practice can provide the location in which researchers can operationalise, test and facilitate sustainability innovations and disruptive experiments. The attitudes of many sustainability transitions researchers transcend the usual boundaries of research involvement, by seeking opportunities to actively influence

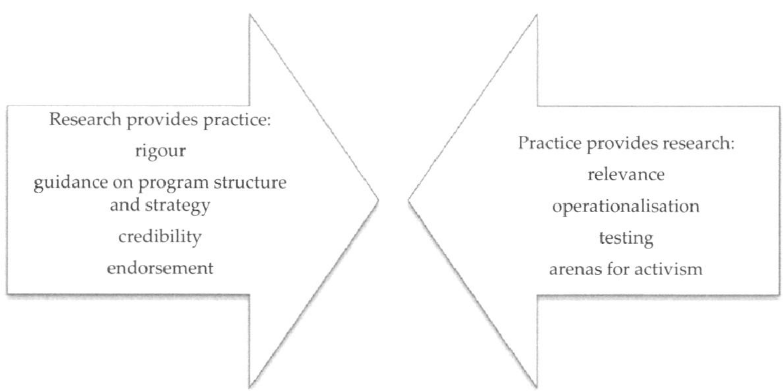

Fig. 5.2 Contributions and benefits of research-practice collaborations

or contribute to these transitions. Loorbach (2014) refers to this as 'activist research', and Jan Rotmans (pers. comm.) coined the term 'scientivist' to represent the integration of the otherwise separate roles of studying transitions and fostering transitions. In this context, practice is more than simply an object of study for research; practice can provide to researchers the arena and vehicle in which they can fill the activist research role, through their active participation in fostering and implementing transitions.

5.5.3 *Implications for Urban Transitions*

Cities and local climate responses are increasingly recognised as significant elements in global efforts to mitigate climate change (Castán Broto and Bulkeley 2013; UN-Habitat 2011). This recognition in part relates to the relative magnitude of urban-related greenhouse gas emissions, as well as the substantial roles that cities, municipalities and community organisations are already playing, sometimes in the face of lack of action at higher levels of government (Bulkeley and Betsill 2013). This research demonstrates that sustainability transition theories can contribute to practical action, through providing validation and reinforcement for niche-level innovations and regime-influencing initiatives. This research also contributes to the ongoing debate of where cities are located within the multilevel perspective (Doyon, this volume). As a result of MEFL's definition of its model of change, it explicitly locates itself in the space between niche and regime, functioning as a bridge or connector between niches, conceived as grassroots organisations and technical innovations, and the regime, conceived as renewable energy policies, regulations and practices (MEFL 2015).

Hodson and Marvin (2010) highlighted the important role that intermediary organisations play as 'agents of change' in bringing together a range of social and governance interests as part of urban transitions. They reinforced the development of capacity and capability of intermediaries as crucial to enabling the translation of sustainability visions and targets to action (Guy et al. 2011; Hodson and Marvin 2010). This research can be understood as demonstrating one approach to this endeavour. By undertaking a research-practice project to translate research theories to practical application, MEFL's capabilities as an intermediary have been strengthened. Its now explicit model of change can contribute to its intermediary efforts to translate its overarching vision and goals to multilevel partnerships and tangible action.

The results of this research are applicable to other intermediary organisations. A number of MEFL participants reflected that the application of the theoretical framework would be useful in other organisational strategic planning processes. 'It made me start thinking about outside MEFL as well, and other people I work with and how it might apply' (Respondent 6). Another commented that they had already discussed the theories with others beyond MEFL, 'someone I know more from the innovation space ... it had a real sense of innovation process. It made a lot of sense, inherently intuitively, to me' (Respondent 2). However, others felt that guidelines and tools, 'templates and useable stuff' (Respondent 6), would need to be developed for the theories to be useful for organisations beyond MEFL. The existing capability of organisations, and their readiness to engage with complex theoretical material, also limits applicability. 'The concern I have is that it is not an easily grasped concept and needs to be quite carefully introduced to a team because it can be quite technical. And I think the MEFL board was ripe for it because we had been grappling with things like doing all this work for so long' (Respondent 5).

Furthermore, application of theories by practice can risk resulting in a loss of nuance and depth and of separation from ongoing theoretical development (as was discussed by Cairney and Jones (2016) in relation to empirical applications of a particular policy theory). If other organisations seek to utilise sustainability transition theories in their strategic planning, 'will they just pick out the bits they like, is that ok?' (Respondent 6). In many ways, these concerns reinforce the strength of partnership approaches, and necessity for maintaining ongoing research-practice relationships, to support continued two-way learning and development of both theoretical concepts and organisational capacity and capability. Future research could examine how productive research-practice relationships are maintained and supported. In addition, assessing the impact of theoretical input on intermediaries' organisational capacity and capability over time is a fruitful area for further research.

5.6 Conclusions

This research has investigated the impact of inclusion of sustainability transition theories into organisational strategic planning processes and the theories' practical application and utilisation beyond the research community. Whilst a range of potential benefits were identified by the researchers at the start of the research process, with the application of grounded theory analysis, the organisational strategic planning participants' own views and responses to the theories emerged, revealing a range of other benefits and contributions.

The inclusion of sustainability transitions theories to MEFL's strategic planning process contributed to both the process and the resulting final strategic plan document. MEFL explicitly defined its model of change as a result of inclusion of sustainability transition theories in the strategic planning process. By defining its model of change, MEFL also reinforced its commitment to working in partnership with other innovators and with business and government. The sustainability transition theories provided MEFL with a new language with which to discuss and elaborate its strategic planning and its approach to implementation and delivery. However, the theories' contribution to strengthening MEFL's organisational visioning, ambition and confidence was less evident in the process, partly due to the relatively short time frame associated with the research input and partly due to MEFL's existing organisational sense of confidence and ambition.

This research has added to understandings of how transition theories can be and are used in practice, as well as how researchers can potentially work with and contribute to strategic planning in practice. It has also contributed to understandings of how theories can be grounded in organisational processes, which in turn could inform further theoretical development.

The research has highlighted the contributions of a research-practice partnership approach to urban sustainability transitions and in particular, the contributions that practice can make to the research community (in addition to simply being an object of study). Practice can inform theory by providing the information and the context or platform to operationalise theory and to test implementation through matching models, concepts and theories with lived experiences. Practice can in addition provide the arena for action research, activist research and 'scientivists' working to foster and facilitate sustainability transitions. Further, transdisciplinary approaches reinforce not just the co-production of data and knowledge but also the initiation of research questions and research approaches, based on practitioner needs as well as research priorities. Transdisciplinarity emphasises a focus on how collaboration happens, maintaining relationships between individuals and research and practice communities and explicitly acknowledging and aiming to complement each other's skills and strengths.

References

ACF (2016) Australia's 10 biggest climate polluters. Australian Conservation Foundation, Carlton

APS (2016) Psychology, climate change and the environment http://wwwpsychologyorgau/public-interest/environment/. Accessed 23 May 2016

Avelino F, Grin J, Pel B, Jhagroe S (2016) The politics of sustainability transitions. J Environ Policy Plan 18(5):557–567

Binder CR, Absenger-Helmli I, Schilling T (2015) The reality of transdisciplinarity: a framework-based self-reflection from science and practice leaders. Sustain Sci 10(4):545–562

Bulkeley H, Betsill MM (2013) Revisiting the urban politics of climate change. Environ Politics 22(1):136–154

Bulkeley H, Castán Broto V, Edwards GAS (2015) An urban politics of climate change: experimentation and the governing of socio-technical transitions. Routledge, London

Cairney P, Jones MD (2016) Kingdon's Multiple Streams Approach: what is the empirical impact of this universal theory? Policy Stud J 44(1):37–58

Castán Broto V, Bulkeley H (2013) A survey of urban climate change experiments in 100 cities. Glob Environ Chang 23(1):92–102

Charmaz K, Belgrave LL (2012) Qualitative interviewing and grounded theory analysis. In: Gubrium JF, Holstein JA, Marvasti AB, KD MK (eds) The SAGE handbook of interview research: the complexity of the craft. SAGE, Thousand Oaks, pp 347–365

CRPA (2016) Health and wellbeing. Climate Reality Project Australia. http://www.climatereality.org.au/health-and-wellbeing.html. Accessed 23 May 2016

Eckersley R (2015) Australian democracy and climate politics for the long term. Meanjin 74(3):140–145

Frantzeskaki N, Kabisch N (2016) Designing a knowledge co-production operating space for urban environmental governance lessons from Rotterdam, Netherlands and Berlin, Germany. Environ Sci Pol. doi:10.1016/j.envsci.2016.01.010

Frantzeskaki N, Wittmayer J, Loorbach D (2014) The role of partnerships in 'realising' urban sustainability in Rotterdam's City ports area, the Netherlands. J Clean Prod 65:406–417
Geels FW (2002) Technological transitions as evolutionary reconfiguration processes: a multi-level perspective and a case-study. Res Policy 31(8–9):1257–1274
Geels FW (2011) The multi-level perspective on sustainability transitions: responses to seven criticisms. Environ Innov Soc Trans 1(1):24–40
Geels FW (2014) Regime resistance against low-carbon transitions: introducing politics and power into the multi-level perspective. Theory, Cult Soc 31(5):21–40
Guy S, Marvin S, Medd W, Moss T (2011) Shaping urban infrastructures: intermediaries and the governance of socio-technical networks. Earthscan, London
Hamilton C (2010) Requiem for a species. Why we resist the truth about climate change. Earthscan, London
Harre N (2011) Psychology for a better world. Department of Psychology/University of Auckland, Auckland
Hes D, Du Plessis C (2015) Designing for hope: pathways to regenerative sustainability. Routledge, Abingdon
Hodson M, Marvin S (2010) Can cities shape socio-technical transitions and how would we know if they were? Res Policy 39(4):477–485
Kent J (2016) Community action and climate change. Taylor and Francis, London
Loorbach D (2010) Transition management for sustainable development: a prescriptive, complexity-based governance framework. Governance 23(1):161–183
Loorbach D (2014) To transition! Governance panarchy in the new transformation. DRIFT, Rotterdam
Markard J, Raven R, Truffer B (2012) Sustainability transitions: an emerging field of research and its prospects. Res Policy 41(6):955–967
McGuirk P, Dowling R, Brennan C, Bulkeley H (2015) Urban carbon governance experiments: the role of Australian local governments. Geogr Res 53(1):39–52
MEFL (2015) Strategic plan 2015–2018. Moreland Energy Foundation Limited, Melbourne
Moloney S, Fünfgeld H (2015) Emergent processes of adaptive capacity building: local government climate change alliances and networks in Melbourne. Urban Climate 14:30–40
Moloney S, Horne R (2015) Low carbon urban transitioning: from local experimentation to urban transformation? Sustainability 7(3):2437–2453
Narberhaus M (2011) Smart CSOs: effective change strategies for the great transition. Five leverage points for civil society organisations. In: Conference background paper: smart CSOs conference, London, 14–15 March 2011
Nevens F, Roorda C (2014) A climate of change: a transition approach for climate neutrality in the city of Ghent (Belgium). Sustain Cities Soc 10:112–121
Nevens F, Frantzeskaki N, Gorissen L, Loorbach D (2013) Urban transition labs: co-creating transformative action for sustainable cities. J Clean Prod 50:111–122
Positive Charge (2016) About Positive Charge. http://wwwpositivechargecomau/about. Accessed 12 Mar 2016
PSC (2016) Fostering emotional engagement with climate change. Psychology for a safe climate. Accessed 23 May 2016
Rotmans J, Kemp R, van Asselt M (2001) More evolution than revolution: transition management in public policy. Foresight 3(1):15–31
Schot J, Geels FW (2008) Strategic niche management and sustainable innovation journeys: theory, findings, research agenda, and policy. Tech Anal Strat Manag 20(5):537–554
Seyfang G, Haxeltine A (2012) Growing grassroots innovations: exploring the role of community-based initiatives in governing sustainable energy transitions. Environ Plann C Gov Policy 30(3):381–400
Shove E, Walker G (2007) CAUTION! Transitions ahead: politics, practice, and sustainable transition management. Environ Plan A 39(4):763–770
UN-Habitat (2011) Cities and climate change: global report on human settlements. United Nations Human Settlements Programme, London

Ward J, Power M (2015) Cleaning up Victoria's power sector: the full social cost of Hazelwood power station, Environment Victoria, Carlton

Wiseman J, Karoly D, Sheko A (2014) Cool Melbourne: towards a sustainable and resilient zero carbon city in a hotter world. In: Whitzman C, Gleeson B, Sheko A (eds) Melbourne: what next?, Research monograph no 1. Melbourne Sustainable Society Institute. The University of Melbourne, Parkville, pp 60–73

Wittmayer JM, Schäpke N, van Steenbergen F, Omann I (2014) Making sense of sustainability transitions locally: how action research contributes to addressing societal challenges. Crit Policy Stud 8(4):465–485

Yin RK (2014) Case study research: design and methods. SAGE Publications, Thousand Oaks

Chapter 6
'Transitions in the Making': The Role of Regional Boundary Organisations in Mobilising Sustainability Transitions Under a Changing Climate

Susie Moloney, Karyn Bosomworth, and Brian Coffey

Abstract Sustainability under a changing climate requires transitioning away from institutionalised processes, norms and cultures that underpin and reproduce unsustainable practices and development. The volume and diversity of actors, and the closeness and density of interactions and interrelationships, make urban transitions complex, contested and dynamic, challenging established management practices, institutions and governance. Therefore, enabling sustainability transitions requires social processes of adaptive, if not transformative, change and learning, facilitated by improved capacities for working across diverse forms of jurisdiction, scale, knowledge, organisations, landscapes and institutions. Recognition of the challenges inherent in these issues has led to arguments for new forms of governance, such as Transition Management. The dynamic relations between niche and regime have been identified as requiring further analytical attention. In our research, we have identified 'boundary organisations' as operating in this space as they work to enable energy and natural resource transitions in Victoria. This paper explores what we are learning about and from these organisations in enabling some of the conditions considered important in the governance of transitions, such as experimentation, long-term thinking and learning by doing across multiple boundaries.

Keywords Sustainability governance • Sustainability transitions • Boundary organisations • Natural resource management • Local governments

S. Moloney (✉) • K. Bosomworth • B. Coffey
Centre for Urban Research, School of Global, Urban and Social Studies, RMIT University, Melbourne, Australia
e-mail: susie.moloney@rmit.edu.au; karyn.bosomworth@rmit.edu.au; brian.coffey@rmit.edu.au

T. Moore et al. (eds.), *Urban Sustainability Transitions*, Theory and Practice of Urban Sustainability Transitions, DOI 10.1007/978-981-10-4792-3_6

6.1 Introduction

In light of major sustainability challenges confronting society including climate change, biodiversity depletion, freshwater and food security and social inequities (Rockstrom et al. 2009), attention is increasingly given to how we might 'steer' or govern transitions towards sustainability. Enabling such transitions requires a nuanced understanding of governance including processes of institutional change and inertia and how political, sociocultural and policy shifts might occur in different contexts. Transitioning is understood as involving non-linear "context-dependent evolutionary processes with emergent properties" (Turnheim et al. 2015: 240). In analysing how or if transitions are actually occurring, insights from a range of analytical approaches are required to better understand how context shapes, enables and disables transitioning. Contributing to the research agenda concerned with sustainability transitions governance, this paper examines the work of 'boundary organisations' – generally understood as formal and informal organisations working across science-policy boundaries in complex governance contexts – and argues that these types of organisations can play an important role in mobilising transitions for two reasons. First, they can bridge boundaries between niche innovation and regime 'inertia', actively working to reconfigure multiple elements of the sociotechnical systems involved in sustainability transitioning. Secondly, they constitute spaces that enable experimentation across jurisdictional and other boundaries, creating openings for learning and reflexivity.

Informed by governance, policy and sustainability transitions-oriented literature, we present a comparative analysis of two different types of boundary organisations operating across regional scales in Victoria, Australia, to progress more effective responses to the challenges of sustainability and climate change: the formally established Catchment Management Authorities (CMAs) who seek to enable transitioning in natural resource management towards social-ecological sustainability across regions of Victoria and voluntary collaborations between local governments ('Climate Change Alliances or CCAs') who seek to enable low-carbon energy and urban transitions in particular regions of Melbourne and Victoria. In our analysis, we first situate our work within the context of sustainability transitions and governance and draw on recent work around boundary organisations emerging from the science and technology studies (STS) literature as a way to explore niche regime dynamics; second, drawing on observational evidence and analysis, we outline our two different boundary organisations highlighting their emergent characteristics in mobilising transitions in Victoria; and third, we discuss insights from these case studies for the theory and practice of analysing and enabling sustainability transitions.

6.2 Governing Sustainability Transitions

Within the context of transition studies as 'attempting to understand and explain the dynamics of fundamental, long-term societal change' (Roorda et al. 2014: 9), the multilevel perspective (MLP) provides a useful conceptual device to examine the evolution of and dynamic relationship between multiple elements and processes of sociotechnical transitioning including changes in user practices, regulation, production and consumption networks and systems, infrastructure and meanings and culture (Geels 2002). Originally focusing on technological transitioning over time, Geels (ibid) contended that the key point of the MLP is that technological transitions occur as outcomes of linkages and interactions of development at multiple levels across landscape-regime-niche and involves the reconfiguration of linkages as well as elements within systems.

In examining sociotechnical systems and transitions, studies are concerned with understanding and explaining how transitions to new sociotechnical systems come about highlighting the importance of niches as the 'locus of radical innovations' that 'act as "incubation rooms" for radical innovation nurturing early development' (Geels 2005: 451). Niches have little stability and are characterised by uncertainty unlike incumbent regimes which constitute elements that are closely aligned. Further Geels (*ibid*) argues that niches provide space 'for learning processes on several dimensions, for example technologies, user preferences, regulations, infrastructures and symbolic meaning' and 'to build social networks that support innovations for example lobby groups, user associations and new industry networks' (451). These 'protected spaces' are viewed as important in enabling experimentation and novel processes to emerge through trial and error and learning. The capacity for niches to 'break through' and shape or transform regimes is central to sustainable transitions assessments. Sociotechnical regimes co-evolve and are shaped and held together by a 'broad constituency' of actors, interests and material elements (Geels and Schot 2007: 479). Analyses of regimes examine the configuration of institutions, practices and regulations 'where configurations impose a logic, regularity and varying degrees of path dependences on technological change' (ibid: 479).

The notion of regime change resulting from emergent niches 'working upwards' is contested, arguing that governance processes involve a dynamic relationship across all levels including changes in landscape pressures and conditions (Smith et al. 2005; Berkhout et al. 2003; Bevir and Rhodes 2001). In the context of how to better understand 'transitions in the making' and processes of innovation and inertia, Turnheim et al. (2015) contend that:

> if transitions are about both the generation of novelty and the destabilisation of incumbent regimes, effective analytical approaches will also need to capture these two sides of the problem. (Turnheim et al. 2015: 241)

In contributing to better frameworks for understanding governance of 'sustainability pathways', Turnheim et al. (2015) also identify three categories of perspectives on governing transitions: *command and control public policy* which focuses on government as the main actor addressing problems of legitimacy and effectiveness; *public-private governance* focusing on government and business as the main actors and issues of incentives, accountability and effectiveness; and *adaptive governance* as responding to 'emergent properties of complex transition processes' focusing on a range of actors including business, civil society and government and the problems of coordination, assessment and intervention (Turnheim et al. 2015: 242). Each approach they argue emphasises specific aspects of transitions while neglecting others and taken separately will be inadequate to address the central challenges presented by sustainability transitions: the multiple scales, geographies and temporalities of transformational processes, uncertainties associated with radical innovation and the limits of prediction, the inertia of sociotechnical systems and emergence of novelty, the need to address multiple normative goals and objectives and contested perspectives on governance (ibid, p240). Turnheim et al. (2015) also argue that while the diversity of perspectives on governance complicates a more integrated analytical approach to sustainability transitions, they contend that analyses should seek to make contributions to all three.

With this in mind we engage with governance issues inherent in sustainability transitions through examining the role of two types of boundary organisations in Victoria that navigate a range of scales, geographies and jurisdictions and across multilevels of landscape-regime-niche dynamics: CCAs and CMAs. We draw on recent STS literature on boundary organisations and suggest that these types of organisations potentially intersect with all three perspectives on governing transitions. In exploring the characteristics of boundary organisations in both the literature and practice through our two case studies, we explore whether and how this concept adds further insight to understanding governance of transitions or 'transitions in the making'.

In exploring how different perspectives on governance are involved in transitioning, we draw also on contributions from 'urban governance' and the politics of transitioning (Hodson and Marvin 2010; Bulkeley et al. 2014; Voss and Bornemann 2011). Hodson and Marvin (2010) draw attention to the necessary mediating roles of 'intermediary organisations and contexts' and to examining the politics of whose priorities are dominant and what the implications are for urban transitions. They argued that 'the creation of intermediaries is necessary to constitute a space outside of the obduracy of both existing urban governance regimes and existing socio-technical regimes' (Hodson and Marvin 2010: 482). They raise the issue, when analysing transitioning in different contexts focusing on the various actors and coalitions involved, whether actors are working towards a 'genuine, radical transition' or just continuing to reproduce the status quo. They contend that researchers need to focus on the 'where' of transitions approaches to better understand the various urban contexts or conditions shaping and mediating transitions and importantly identify to what extent sociotechnical systems and their transition can be governed and configured at the urban scale. We contend here that boundaries between urban and 'non-urban' and what constitutes urban governance also need careful consideration.

Associated with the notion of intermediary organisations as important in operating across multiple scales and creating spaces outside sociotechnical regimes is the concept of adaptive governance, which we view as being concerned with transformative rather than incremental change (Chaffin et al. 2014). In the context of climate change adaptation, studies of adaptive governance systems have emerged in recognition that transformative processes require changing existing institutions, organisational arrangements and practices. Similar to the question of whether intermediaries are pursuing radical transitions rather than reproducing the 'status quo', Pelling and High (2005:309) highlight the value of differentiating actions that 'reinforce existing organisational or system stability and those that modify institutions to add resilience through flexibility'. Processes of social and societal learning, important features of adaptive governance, are necessary to respond to the complex challenges of resources management and climate change (Pahl-Wostl 2009; Folke et al. 2005). It is contended that adaptive and reflexive modes of governance aim to move away from a 'command and control' paradigm to one that constitutes more active, deliberative involvement from a range of stakeholders in the design and management of policies and plans (Pahl-Wostl 2009: 354; Voss et al. 2006). In the same way that analyses in transitions studies are called to better understand both the 'generation of novelty' and the 'destabilisation of incumbent regimes' (Turnheim et al. 2015: 241), studies of informal adaptive networks are important particularly in terms of both their emergent characteristics and their capacity to change existing regimes. These types of networks can be described as 'self-organising groups of policymakers' who are influential in and have knowledge about different power networks but, importantly, 'try to break away from the existing policies in those power networks and develop joint understanding about new, more effective policies in these informal adaptive networks' (ibid: 361). The boundaries between what constitutes formal and informal in governance arrangements and processes are not always clear, nor the boundary between niche and regime – these are issues we explore further in our two case studies.

6.3 Boundaries and Boundary Organisations

While this book focuses on 'urban transitions', we recognise that distinguishing between 'urban' and 'natural' boundaries is not always clear nor useful when considering sustainability transitions. This point is well made by Rickards et al. (2016) who argue that notions of the city and the urban as self-evident policy lenses are too static and too bounded. We argue that it is important to critically engage with and explore boundaries within governance, whether spatial, knowledge or organisational to better inform the theory and practice of what it means to enable sustainability transitions. Governing 'urban transitions', for instance, implies a spatial and jurisdictional boundary that may be inappropriate in the context of climate change, given that 'the urban' is situated within broader biophysical and social landscapes. Consider the locations that provide the sources of water and food for urban areas. For example, in Victoria, CMA boundaries broadly align with water catchment boundaries. As such, they have to navigate urban/rural boundaries, as well as multiple social-ecological system and jurisdictional boundaries. Informal networks and alliances of local governments concerned with mobilising local- and regional-scale climate change governance also contend with multiple spatial (urban-rural, inner-middle-outer urban, etc.), infrastructural (energy, water, transport) systems and jurisdictional (cross local government) boundaries. Transitioning to a low-carbon metropolitan Melbourne, for example, will necessitate significant economic and social transitioning in the regional area of the Latrobe Valley, where most of the State's electricity is produced by burning brown coal. Enabling sustainability transitions therefore requires consideration of how system boundaries are understood, navigated, managed and governed across including sociotechnical, social-ecological, science-policy and organisational boundaries. Issues associated with the importance of place in transitions are explored in Chap. 2 of this book.

Here we briefly examine the concept of boundary organisations that has its origins in the STS literature, with emerging studies exploring their role in governance of climate change and natural resource management (NRM) issues. According to Hoppe and Wesselink (2014: 74), 'boundary organisations were conceived by Guston (2001) as organisations that "straddle the shifting divide between politics and science"' and are embedded in and shaped by their contexts of networks and political-cultural spheres, which help to facilitate productive interactions between science, policy and politics (ibid: 74). Whether formalised or not, they are 'sites of what Sheila Jasanoff (1996: 397), following Bruno Latour, has labelled "coproduction": the simultaneous production of knowledge and social order' (Guston 2001: 401). Clark et al. (2011) highlight that this boundary work 'through which research communities organize their relations with new science, other sources of knowledge, and the worlds of action and policymaking' broadly involves six types of boundary work:

- Demarcation of knowledge
- Provision of expert advice
- Assessment of advice
- Integrative research and development
- Participatory research and development
- Political bargaining

They argue that the breadth of boundary work presents two challenges for boundary workers: the need to integrate multiple forms of knowledge, in particular the contextualised knowledge of practice with the generalised knowledge emerging from research, and the politicisation of formal knowledge, including the relationship between knowledge and power.

In this chapter, we argue that CMAs and CCAs not only work across these sorts of research, policy and practice boundaries but that they also work across multiple other forms of boundaries. These organisations work in contexts in which new research and practice insights from multiple policy domains have to intersect with established (and currently applied) knowledge, and they must integrate information from a range of research disciplines (and researchers) *alongside* integrating a range of information from a variety of policy and political domains. They are also sites of research application (again, often integrating research from a variety of sources/disciplines/policy arenas) and 'testing' in terms of policy relevance.

We contend in this paper that as well as dealing with science-policy boundaries, these organisations also work across boundaries between local/territorial (a particular river or location within a municipality) and strategic (whole-of-catchment or regional) scales. Further, they work across spatial, temporal, rural, urban, ecological, political, social and economic boundaries. From a Transition Management perspective (see Loorbach 2010), these organisations take their local stakeholder and collective experiences back into the regime (back into policy). It is because CMAs and CCAs negotiate or facilitate across multiple boundaries that we argue are boundary organisations. It is this capacity for/role in negotiating across and between differing values and knowledge that ideally places such boundary organisations to support social processes of transformative change required in sustainability transitions.

In the context of developing further insights into 'transitions in the making' and processes of governance, we argue that boundary organisations are likely playing an important role and further analysis of that role and the boundaries they navigate is warranted. We propose that while boundary organisations may not constitute 'transition arenas' incorporating a network of 'frontrunners' (Loorbach and Rotmans 2010: 237), in considering the multiplicity of boundaries they navigate, these types of organisations certainly demonstrate similar activities and may in fact be important actors in a transitions arena where one is to emerge or be created. In the context of the MLP, boundary organisations can be characterised as operating across the boundary between niche regime and certainly intersecting with and responding to landscape changes. Further, while boundary organisations may not constitute 'niches', they may create the space for niche experiments to emerge. While these types of organisations reflect similar ideas of mediation explored in sociotechnical literature, the notion of 'boundary work' as explored in STS literature potentially offers further insights into mediation work particularly across science-policy boundaries and other boundaries of various sorts in processes of transitioning. Leith et al. (2015: 375) also challenge 'the idea that the success of boundary organisations is marked primarily by the stability of the science-policy interface' and instead argue that their success is indicated 'not just through their surviving periodic controversies, but by being able to benefit from them, building legitimacy among partners and stakeholders through successfully navigating unforeseen events' (ibid: 376). Given this, we consider that the concept of boundary organisations – as conceived in this paper – can make a useful contribution to enabling Transition Management because they can exhibit a capacity to thrive in contested situations (political cycles and shifts in policy emphases), challenging the regime and enabling niche experiments. Put simply, they provide a space for negotiating and redefining the boundaries between regime and niche.

6.4 Analysing Boundary Organisations in Sustainability Transitioning

We now briefly outline the context – the state of Victoria – as a useful case study of sustainability transition efforts and present an overview of the two boundary organisations examined – Catchment Management Authorities and Climate Change Alliances – who seek to enable transitions in natural resource management and low-carbon transitions, respectively. We include an outline of what they are and how they have emerged, the type of work they do and how they do it and the challenges they confront as their role and work evolves. In the discussion section, we draw on insights from each case to explore the role and work of boundary organisations in enabling transitions and what these insights offer to further develop our understanding of the governance of sustainability transitions in theory and in practice.

6.5 Victoria as a Case Study

Victoria is a suitable case for exploring the governance of sustainability transitions for three reasons. First, under Australia's Constitution, environmental matters and arrangements for local government are primarily a state government responsibility, as the national government has no explicit constitutional powers in these areas, although it can use financial and indirect constitutional powers to exert influence (e.g. external affairs power). Under these arrangements, there is no metropolitan or regional scale of government, and local government responsibilities are defined at the state level. Second, Victoria has a reputation within Australia for being a policy innovator (Nelson 1985), with examples from the environmental domain including the whole environment focus of its Environment Protection Act (EPA 1996), the use of a statutory authority model for strategic public land use planning (Coffey et al. 2011), the establishment of a Commissioner for Environmental Sustainability (CES) (State Government of Victoria 2000) and an innovative approach to flora and fauna protection (Walker 2003). Third, Victoria is facing considerable sustainability challenges (VCMC 2012, CES 2013). It is the second smallest Australian state in the area (22.76 million hectares) yet is the second most populated state (Wescott 1998), with a population of 5.53 million people, of which 75% live in the Greater Melbourne area (CES 2013). Overarching drivers of environmental degradation which have been identified are population growth, the use of energy and water and consumption and waste (CES 2013: 192), while more than half of the state has been cleared of native vegetation (CES 2008). Climate science also indicates that Victoria faces a warmer and drier future, resulting in harsher fire weather and longer fire seasons; fewer frosts; more frequent and more intense downpours; more hot days and warm spells; less rainfall in winter and spring south of the Great Dividing Range; less rainfall in autumn, winter and spring north of the Great Dividing Range; and sea storm surges and coastal erosion that are expected to increase with sea level rise (DELWP 2015). Such evidence demonstrates that sustainability transitions are necessary.

6.6 Climate Change Alliances (CCAs)

While other Australian states include a number of partnerships and local government associations, Victoria is unique in having 72 of 79 local governments involved as voluntary members in regional Climate Change Alliances (CCAs) or partnerships. Of a total of ten alliances, four are most active and cover all but six metropolitan councils in Melbourne and are the focus on this paper. These include the Western Alliance for Greenhouse Action (WAGA), Northern Alliance for Greenhouse Action (NAGA), Eastern Greenhouse Alliance (EAGA) and South Eastern Councils Climate Change Alliance (SECCCA). Each alliance crosses multiple adjacent local government boundaries including four or more councils who each contribute a small membership fee annually to support their work. Spatially these alliances incorporate inner-middle and outer suburbs including both urban/peri-urban and rural landscapes. The metropolitan alliances tend to be more adequately resourced and perhaps, because of proximity (not covering such large geographical areas) and a longer period of operation in most cases, have been more successful in building their regional capacity and in supporting member councils to respond to climate change.

In the early 2000s, the alliances emerged as informal 'information sharing' networks amongst local government sustainability officers. In 2002 the Victorian Labor government established the 'Regional Partnerships Program' as part of its Victorian Greenhouse Strategy. The government recognised there was a need to improve the capacity at the local scale through supporting networks and partnerships. The programme funded an executive officer for each alliance with initial work focusing on internal council climate change mitigation initiatives. In the mid- to late 2000s, alliance work focussed on implementing energy-efficient street lighting and council buildings. In 2006, the Regional Partnerships Program was reviewed by the state government, and the importance of alliances in facilitating capacity building and progressing mitigation actions across councils was highlighted. Nevertheless, the state government ceased its funding support of a chief executive for each alliance in 2008 leaving alliances to operate entirely independent of the state government with ongoing contributions from their member local governments. They operate on limited funds maintaining a chief executive, and through applying for grants they build their capacity and staff to develop regional strategies and implement projects.

Each alliance typically is constituted by an executive committee and working group including representatives from each of their member councils.

While alliances are not formally recognised by the state government, they exist largely because they continue to provide ongoing strategic support around climate change action for their member local governments. Over time the CCAs have managed to maintain and strengthen their role in delivering and implementing a range of regional and local climate change strategies and projects despite an often hostile climate change policy context at both the state and federal levels. While the federal government has played varying roles over the years in establishing/dismantling emissions reduction targets and the state government in developing/dismantling climate change policy and legislative frameworks, the lack of stability and inconsistency through electoral cycles means local governments have been required to manage the impacts and challenges of climate change with limited political or financial support. The CCAs have worked to build capacities across their councils, and in forming a 'coalition' of alliances, they have also sought to build their own capacities to work across regional scales (across multiple municipal boundaries) including infrastructural and urban/rural boundaries.

6.7 Catchment Management Authorities (CMAs)

Victoria's approach to natural resource management NRM is guided by the concept of integrated catchment management (ICM), which involves 'regional, holistic approaches to environmental decision making' (Ewing 2003: 393). ICM is implemented through Victoria's catchment management framework (DELWP 2016a, b) centred around a statewide council – responsible for the preparation of a statewide catchment condition report – and ten regionally based CMAs responsible for the development and oversight of Regional Catchment Strategies (RCS). The Catchment and Land Protection Act (CALP Act 1994) and the Water Act (1989) are the two main legislative instruments related to catchment management. The primary statutory responsibility of each CMA is to develop and implement a 5-year RCS that identifies priorities and targets for managing the region's environment and natural resources. CMAs are also responsible for waterway management and regional drainage and are the custodians of the Environmental Water Reserve.

The CALP Act requires that the council and CMAs have a community-based membership, broadly focussed on 'people working better together to coordinate planning, investment and on-ground activities as a more effective and efficient way to achieve a range of environmental, economic, and social outcomes' (DELWP 2016b: 14). Each CMA is governed by a board, which sets regional strategic priorities, evaluates the effectiveness of outcomes, monitors external and internal CMA environment and identifies strategic opportunities. The board and CMA staff are responsible for developing and implementing programmes and for liaising with communities, government and other organisations (DELWP 2016a).

The Victorian Government recently reinvigorated Victoria's catchment management framework (DELWP 2016b) as climate change mitigation and adaption becomes increasingly central in efforts to create and sustain climate-smart and resilient landscapes, communities and industries. For example, plans for dealing with the NRM aspects of climate change are being developed during 2016 (DELWP 2016b: 19). Development of these plans was recently funded, over 2 years, via the national government. Planning for and acting on climate change represent both a considerable challenge and opportunity for CMAs, whose origins were quite specifically focussed on soil conservation and regional river improvement trusts (DELWP 2016b: 10).

Climate change is a 'game changer' in NRM. Looking after land, water and the other species with which we share landscapes underpins sustainability, and without addressing widespread trends of continuing degradation, we are unlikely to pass on the country to future generations in a better condition than we found it (Wallis et al. 2015). Limiting degrading processes and pressures, let alone maintaining or enhancing the condition of our natural resources, will be an increasing challenge under a changing climate (*ibid*). The preservation or 'stationary' basis of much NRM planning and management (Milly et al. 2007) is no longer feasible or relevant under a changing climate because of the pervasive and large changes anticipated under significant levels of climate change (Dunlop et al. 2013: 2). At the same time, interest in sequestering carbon through various land management actions receives growing attention. Consequently, the CMAs' focus has expanded from the implications of climate change for NRM and the need for adaptation, to include carbon sequestration that supports adaptation efforts. The fundamental challenge is how to support transitions from traditionally productivist landscapes (unsustainable land uses in regional areas; marginal, piecemeal efforts) to multifunctioning, multipurpose landscapes (socially ecologically sustainable landscapes). Moreover, CMAs need to be able to justify, prioritise and implement actions while at the same time recognising and allowing for future changes in ecosystems, values, knowledge as well as sociopolitical and climatic systems (Bosomworth et al. 2015).

6.8 Analysis: The Role of Regional Boundary Organisations in Mobilising Sustainability Transitions Under a Changing Climate

This analysis seeks in part to address Turnheim et al.'s (2015: 241) contention that 'if transitions are about both the generation of novelty and the destabilisation of incumbent regimes, effective analytical approaches will also need to capture these two sides of the problem', paying attention to the role of different forms of governance (i.e. command and control public policy, public-private governance and adaptive governance). Through this analysis, we argue that transitions may not always be about destabilisation of the regime but rather that points/moments of change can come from both inside the regime (CMAs as shakers/shifting regime) and outside (CCAs as creating pressure on regime). In the following we highlight how boundary organisations intersect with the three forms of governance outlined by Turnheim et al. (2015) and explore first how they are enabling the reconfiguration of multiple elements of sociotechnical systems involved in sustainability transitioning by bridging niche innovation and regime 'inertia' and second how they constitute spaces that enable experimentation across jurisdictional and other boundaries creating openings for learning and reflexivity.

6.9 Bridging Niche Innovation and Regime Inertia

CMAs are formally constituted through legislation that embeds their function and enables their continuity across changes in government and related legislation. CMAs and the broader catchment management framework are recognised as a robust and innovative approach to environmental governance (Ewing 2003; Bolitho and Coffey 2014). Their formalised structure and legislated function both require and enable CMAs to work across a number of boundaries, including economic, social and environmental; land, water and biodiversity; urban-rural; public and private land; diverse land uses and users; and public, private and community sectors. The number of boundaries they work across is reflected in the diversity of partners interested in catchment management, including aboriginal communities, agricultural industries, the national government, local governments, individual landholders, conservation groups, universities, various state government departments covering economic development and environmental portfolios, state government statutory bodies and associated interest groups (DELWP 2016b). In working across and negotiating between this multitude of boundaries, CMAs intersect with all three forms of governance. The formal establishment but broad operation of CMAs arguably means they are part of the 'regime', yet their flexibility in performing their formal role represents an innovation established by the regime to manage particular issues. For example, as waterway and floodplain managers (except within the Port Philip and Westernport region), CMAs exhibit a form of command and control governance in setting rules for land use planning. Under the state's Water Act of 1989, CMAs have management powers over regional waterways, floodplains, drainage and environmental water that are expressed through a statement of obligations (DELWP 2016b). While most of these obligations are negotiative, planning and guiding in nature, CMAs also authorise works and activities on waterways and act as a referral body for planning applications, licences to take and use water and construct dams and other waterway management issues (DELWP March 2016b).

Equally, CMAs operate to achieve many sustainability goals through multiple public/private partnerships that fund private action for both the private and public good. Victoria's natural resource management (NRM) is targeted towards river catchment management on private land enabled through a number of community-based conservation programmes, including Land for Wildlife, Farm Tree Groups,

Australian Trust for Conservation Volunteers and the Landcare Programme. Broadly speaking, Landcare is a community-based programme that seeks to fund activities that build community capacity, develop partnerships and engage landholders, NRM community groups, networks and their volunteers in undertaking effective on-ground works to increase ecosystem resilience and contribute to the sustainability of landscapes. Many of the activities undertaken by CMAs also involve considerable financial and other support from individual land managers.

The range of relationships and networks established over time with CMAs emerged to manage NRM issues and demonstrate forms of adaptive governance. Each group or network can focus on different issues and establish forms of management and function according to their needs, actively engaging in research and working between multiple levels of government along with business and community groups.

CCAs are a form of adaptive governance (Moloney and Fuenfgeld 2015) but also seek to influence and shape the 'command and control' functions of the local and state government particularly around regulations affecting energy infrastructure, energy efficiency and urban planning and land use. For example, CCAs played an important role in advocating for changed regulations to allow all local governments to facilitate 'Environmental Upgrade Agreements' (a finance mechanism to encourage private sector commercial building owners to carry out energy efficiency measures). CCAs also form partnerships with a range of stakeholders across the public and private sectors further developing the capacity of local governments to act as facilitators and advocates for change. For example, the Western Alliance for Greenhouse Action (WAGA) is working in partnership with a range of public and private sector partners in implementing 'Low Carbon West' the first regional-scale low-carbon economic development plan in Victoria. The South East Councils for Climate Change Action (SECCCA) operates as an 'incorporated' entity allowing it to support councils while also developing its entrepreneurial capacities in responding to climate change.

6.10 Creating Spaces to Experiment Across Boundaries

One of the central challenges presented by sustainability transitions is the multiple scales, geographies and temporalities of transformational processes. The MLP framework offers a useful approach to analysing landscape-regime-niche dynamics over time including the particular attention given to the processes of regime destabilisation and the emergence of bottom-up novelty. This approach however does not necessarily illuminate the complexity and multiplicity of boundaries navigated through processes of transitioning nor reveal the challenges involved in navigating, for example, science-policy, jurisdictional, spatial and other boundaries.

CMAs operate across a wide range of spatial, jurisdictional and science-policy boundaries and across different levels of government and government departments to both address policy and legislative demands and also to seek multiple sources of information and to share and learn from each other. Increasingly CMAs are operating collectively as a statewide constituency which increases its capacity to shape and pressure the regime. In bringing together the localised, individual landowner perspectives with those of local, state and even federal governments, CMAs are opening up opportunities to reframe significant natural resource challenges such as tensions between agricultural production and conservation in a time of climate change.

CCAs operate across multiple local government jurisdictions as well as between different levels of government local, state and federal. For example, CCAs have been successful at leveraging state and federal funding to implement regional-scale climate change actions. This role has been important in legitimising their value to member local governments (Moloney and Horne 2015). CCAs, while initially working across sustainability departments in local governments, are now working across multiple departments, sectors and agencies as they address climate change mitigation and adaptation challenges at the local and regional scale. This capacity to facilitate cross jurisdictional and cross sectoral collaborations is challenging but important in sustainability transitioning. As boundary organisations, CCAs also navigate different organisations particularly responsible for infrastructure provision and management which do not align with local government boundaries, for example, in working with energy distributors and providers.

In seeking to pursue a range of climate change strategies and actions, CCAs also navigate different spatial boundaries including different urban areas (i.e. inner, middle and outer suburbs), peri-urban regions and rural and coastal areas – each presenting different types of challenges for sustainability transitioning. Challenges include inner urban intensification to outer fringe urban sprawl and inadequate infrastructure, as well as councils with different political leanings, goals, resources and capacities.

CCAs create 'protected spaces' or opportunities to generate new knowledge and information to inform local government action, for example, by commissioning research and facilitating spaces for learning across members and agencies. This necessarily involves navigating the science-policy boundary where, for instance, specific knowledge around the localised impacts of climate change is required. The alliances also work collectively to create a space for learning around different experiments that each may trial, where lessons are used to improve future actions and inform state-level decision-makers. The capacity for alliances to influence the 'command and control' functions of the state government is emerging and will likely improve as they build the evidence around effective actions and strategies.

6.11 Conclusion

This chapter contributes to the research agenda concerning sustainability transitions governance, through examining the work of 'boundary organisations' – formal and informal organisations working across a multiplicity of boundaries including science policy, spatial, jurisdictional and sectoral. Because they work across a range of boundaries, these types of organisations can and are playing an important role in mobilising transitions in natural resource management and low-carbon transitioning. This argument is evidenced through the two cases presented in this chapter – Catchment Management Authorities and local government Climate Change Alliances in Victoria – that can be understood as constituting niche spaces to enable experimentation across jurisdictional and other boundaries creating openings for learning and reflexivity across a range of organisations in addressing NRM and climate change issues in different geographical contexts. As organisations that bridge the boundaries between niche innovation and regime 'inertia', these organisations can and do actively work to enable the reconfiguration of multiple elements of the sociotechnical systems involved in sustainability transitioning.

We contend that further research and practice around the role and work of boundary organisations offer considerable potential to better understand 'transitions in the making', particularly the importance of working with and across diverse geographical boundaries, forms of knowledge, formal and informal organisations and scales of governance involved in sustainability transitioning. This analysis shows that boundary organisations (as conceptualised here) can contribute to building local capacities and transformative shifts in governance that support transitions towards sustainability and outline factors influencing this potential. It contributes to governance thinking in relation to sustainability transitions, highlighting challenges and tensions in working across boundaries including spatial, urban/rural, scale, jurisdictions and even different governance modes. The governance arrangements and practices exhibited by the case study boundary organisations create the necessary spaces for experimentation and learning and are actively working to translate that learning back into the regime to guide or enable transitioning. They seek to shift institutional cultures and reframe problems particularly pushing for long-term planning in the context of climate change. Equally, their arrangements can present challenges around maintaining capacity, longevity and renewal over the longer term, and drawing on lessons from Transition Management approaches may help guide their work in the future. This analysis also suggests there is no one-size-fits-all approach to the design, establishment and operation of boundary organisations; they need to be 'fit for purpose' – the right organisation, in the right place, at the right time. While many questions remain, we posit that processes of transitioning are occurring in Victoria in different contexts and scales and that the kinds of boundary organisations examined herein are playing important roles in these processes. Despite political and institutional obduracies in the climate change and the sustainability space, the studied boundary organisations are contributing to shifting institutional cultures and reframing problems. Some emerging questions for future research focus around issues of politics and power within different modes of governance and the 'operation' of boundary organisations. For example, what are the changing locus of agency and power in different contexts, networks, and decision-making processes, and how effective emerging modes of governance are in shifting political priorities and agendas towards sustainable outcomes? What modes of governance or combinations thereof are needed for different contexts and different issues? Recent research around the role of different approaches to governance at different stages of transitioning is valuable (Rijke et al. 2013) and warrants further exploration in relation to boundary organisations. Finally, how can better understandings of 'transitions in the making' in different contexts help inform or enable the governance of sustainability transitions?

References

Berkhout F et al (2003) Negotiating environmental change: new perspectives from social science. Edward Elgar, Cheltenham/Northampton

Bevir M, Rhodes RA (2001) A decentered theory of governance: rational choice, institutionalism, and interpretation. Institute of Governmental Studies, Berkeley

Bolitho A, Coffey B (2014) Twenty years of integrated catchment management in Victoria: celebrating the achievements of the catchment and land protection act (1994) and looking to the future. State of Victoria/Victorian Catchment Management Council, Melbourne

Bosomworth K, Harwood A, Leith P, Wallis P (2015) Adaptation pathways: a playbook for developing options for climate change adaptation in natural resource management. Southern Slopes Climate Change Adaptation Research Partnership (SCARP): RMIT University, University of Tasmania, and Monash University

Bulkeley H, Castan-Broto V, Edwards G (2014) The urban politics of climate change: experimentation and the government of socio-technical transitions. Routledge, London

Catchment and Land Protection Act 1994 (Vic) State Government of Victoria, Melbourne, http://bit.ly/2vNUWDt

CES (2008) Victoria state of environment report: living well within our environment. Commissioner for Environmental Sustainability, Melbourne

CES (2013) *State of the environment report*. State of Victoria/Commissioner for Environmental Sustainability, Melbourne

Chaffin B, Gosnell H, Cosens B (2014) A decade of adaptive governance scholarship: synthesis and future directions. Ecol Soc 19(3):56 [Online]

Clark WC, Tomich TP, van Noordwijk M, Guston D, Catacutan D, Dickson NM, McNie E 2011 Boundary work for sustainable development: Natural resource management at the Consultative Group on International Agricultural Research (CGIAR). Proceedings of the National Academy of Sciences, (early view). 200900231

Coffey B, Fitzsimons J, Gormley R (2011) Strategic public land use planning in Victoria, Australia: forty years of trailblazing but where to from here? Land Use Policy 28:306–313

Department of Environment, Land Water and Planning (2015) Climate-ready Victoria. Victorian Government, Melbourne, www.climatechange.vic.gov.au. Accessed 5 June 2017

Department of Environment, Land Water and Planning (2016a) http://delwp.vic.gov.au/water/governing-water-resources/catchment-management-authorities. Accessed via www.delwp.vic.gov.au, Mar 2016

Department of Environment, Land, Water and Planning (DELWP) (2016b) Our catchments our communities: integrated catchment management in Victoria 2016–19. Victorian Government, Melbourne

Dunlop M, Parris H, Ryan P, Kroon F (2013) Climate-ready conservation objectives: a scoping study. National Climate Change Adaptation Research Facility (NCCARF) Gold Coast

EPA (1996) The environment protection act 1970: 25 years of making a difference. Environment Protection Authority, Melbourne

Ewing S (2003) Catchment management arrangements. In: Dovers S, Wild RS (eds) Managing Australia's environment. The Federation Press, Sydney, pp 393–412

Folke C, Hahn T, Olsson P, Norberg J (2005) Adaptive governance of social-ecological systems. Annu Rev Environ Resour 30:441–473
Geels F (2002) Technological transitions as evolutionary reconfiguration processes: a multi-level perspective and case study. Res Policy 31:1257–1274
Geels F (2005) The dynamics of transitions in socio-technical systems. Tech Anal Strat Manag 17(4):445–476
Geels FW, Schot JW (2007) Typology of sociotechnical transition pathways. Res Policy 36(3): 399–417
Guston D (2001) Boundary organizations in environmental policy and science: an introduction. Sci Technol Hum Values 26:399–408
Hodson M, Marvin S (2010) Can cities shape socio-technical transitions and how would we know if they were? Res Policy 39:377–485
Hoppe R, Wesselink A (2014) Comparing the role of boundary organizations in the governance of climate change in three EU member states. Environ Sci Pol 44:73–85
Leith P, Haward M, Rees C, Ogier E (2015) Success and evolution of a boundary organisation, science. Technol Hum Values 41(3):375–401
Loorbach D (2010) Transition management for sustainable development: a prescriptive, complexity-based governance framework. Gov An Int J Policy Adm Inst 23(1):161–183
Loorbach D, Rotmans J (2010) The practice of transition management: examples and lessons from four distinct cases. Futures 42(3):237–246
Milly PCD, Julio B, Malin F, Robert M, Zbigniew W, Dennis P, Ronald J (2007) Stationarity is dead: whither water management? Science 319(5863):573–574
Moloney S, Fuenfgeld H (2015) Emergent processes of adaptive capacity building: local government climate change alliances and networks in Melbourne. Urban Climate 14:30–40
Moloney S, Horne R (2015) Low carbon urban transitions: from local experimentation to urban transformation? Sustainability 7(3):2437–2453. doi:10.3390/su7032437
Nelson H (1985) Policy innovation in the Australian states. Politics 20:77–88
Pahl-Wostl C (2009) A conceptual framework for analysing adaptive capacity learning processes in resource governance regimes. Glob Environ Chang 19:354–365
Pelling M, High C (2005) Social learning and adaptation to climate change. Benfield Hazard Research Centre, Benfield
Rickards L, Gleeson B, O'Callaghan C, Boyle M (2016) 'The Urban': a concept under stress in an interconnected world, The Conversation, 27 June
Rijke J, Farrelly M, Brown R, Zevenbergen C (2013) Configuring transformative governance to enhance resilient urban water systems. Environ Sci Pol 25:62–72
Rockström J, Steffen WL, Noone K, Persson Å, Chapin FS III, Lambin E, Lenton TM, Scheffer M, Folke C, Schellnhuber HJ, Nykvist B (2009) Planetary boundaries: exploring the safe operating space for humanity. *Ecol Soc* 14(2):Art 32. http://www.ecologyandsociety.org/vol14/iss2/art32/
Roorda C, Wittmayer J, Henneman P, van Steenbergen F, Frankzeskaki N, Loorbach D (2014) Transition management in the urban context: guidance manual. Erasmus University Rotterdam, Rotterdam, DRIFT
Smith, A, Stirling A, Berkhout F. (2005) The governance of sustainable socio-technical transitions, Res Policy 34:1491–1510.
State Government of Victoria (2000) *Commissioner for Ecologically Sustainable Development Consultation Paper*. Department of Natural Resources and Environment, Melbourne
Turnheim B, Berkhout F, Geels F, Hof A, McMeekin A, Nykvist B, Van Vuuren D (2015) Evaluating sustainability transition pathways: bridging analytical approaches to address governance challenges. Glob Environ Chang 35:239–253
VCMC (2012) Catchment condition and management report. State of Victoria/Victorian Catchment Management Council, Melbourne
Voss JP, Bornemann B (2011) The politics of reflexive governance: challenges for designing adaptive management and transition management. Ecol Soc 16(2):9 [online]
Voss JP, Bauknecht D, Kemp R (eds) (2006) Reflexive governance for sustainable development. Edward Elgar Publishing, Cheltenham

Walker A (2003) The Victorian Flora and Fauna guarantee act – a toothless tiger quoll? Vic Nat 120:224–237

Wallis PJ, Harwood A, Leith P, Hamilton L, Bosomworth K, Turner SL, Harris RMB, Bridle K (2015) Southern slopes information portal report: climate change adaptation information for natural resource planning and implementation. Southern Slopes Climate Change Adaptation Research Partnership (SCARP), Monash University, University of Tasmania, RMIT University

Water Act 1989 (Vic) State Government of Victoria, Melbourne, http://bit.ly/2jfJuio

Wescott G (1998) Reforming coastal management to improve community participation and integration in Victoria, Australia. *Coast Manag* 26:3–15

Chapter 7
Strategic Niche Management and the Challenge of Successful Outcomes

Trivess Moore

Abstract Within the transitions study literature, strategic niche management is proposed as a practical approach to developing and upscaling sustainability niche development. It posits that key actors within governments have important roles to play in creating protected spaces to allow for niches to establish and position themselves to challenge the existing regime. However, the provision of protected space or the involvement of governments does not guarantee a niche will be successful. Sustainability transitions researchers are now asking why some niche projects are more successful than others, both in their establishment and influences on the broader regime. This chapter explores the sustainable housing niche through three cases from Australia where proactive policy levers have been used to try and accelerate: (a) uptake of residential solar photovoltaics (successful), (b) sustainable housing ceiling insulation retrofit (unsuccessful), and (c) sustainable high-density development (limited success). The focus of the chapter is on assessing why there were different outcomes, including influencing the broader regime and societal change, and what the lessons for government involvement in strategic niche management are. The case studies show that when strategic niche management actions align with socio-technical landscape dynamics, then public sector actors have a greater chance at stimulating regime transformation. Inversely, when the public sector fails to align its actions with landscape forces, outcomes of niche experiments are compromised. Within this, context windows of opportunity are critical to improving the likelihood of success for niches and to ensure they can influence the broader regimes. There are opportunities for government actors to improve the likelihood of successful strategic niche management outcomes if they better integrate these learnings into their development and/or support of niches.

Keywords Strategic niche management • Photovoltaics • Ceiling insulation • High-density development • Government actors • Sustainable housing

T. Moore (✉)
School of Property, Construction and Project Management, RMIT University, Melbourne, Australia
e-mail: trivess.moore@rmit.edu.au

T. Moore et al. (eds.), *Urban Sustainability Transitions*, Theory and Practice of Urban Sustainability Transitions, DOI 10.1007/978-981-10-4792-3_7

7.1 Introduction

Current minimum building standards for the residential sector in Australia, and many countries, fall significantly short of what is required for a sustainable low-carbon future (Moore 2014, see also Chap. 11 by Dalton). However, there are increasing examples around the world of how various actors outside the mainstream can, and are, delivering housing which is more environmentally sustainable, while also more affordable and liveable. The question is how to scale up this sustainable housing niche, especially since the examples are typically limited to relatively isolated applications. One potential way to accelerate the development of this niche is through strategic niche management. To explore this option, I apply a broad definition of niche in this paper, covering the sustainable housing actors (e.g., technology suppliers), the innovation, and the subsidies, as discussed in the case studies below.

There is significant discussion within the sustainability transitions literature around strategic niche management as an approach for providing protection to niches (Smith and Raven 2012; Coenen et al. 2010). Creating a protected space aims to help the niche develop rules, expectations, and stability to test and evaluate new alternatives and to challenge the incumbent regime (Geels and Raven 2006; Geels 2011; Rotmans et al. 2001; Smith and Raven 2012). Strategic niche management proponents argue that this protection can be provided through a "carrot and stick" approach, i.e., a combination of more traditional command and control measures and market-based incentives such as rebates or tax exemptions to influence changes in actor behavior (Caniëls and Romijn 2006).

Researchers such as Smith and others (Smith and Raven 2012; Smith et al. 2014) have identified key elements for constructing such protected spaces. These authors discuss the important requirements of shielding, nurturing, empowering, and developing narratives. Shielding and nurturing provide a protected space for the niche to grow on a more level playing field, for example, point of sale rebates for residential solar photovoltaics. Empowering focuses attention onto whether the niche either "fits and conforms" with the existing regime (e.g., solar photovoltaic rebates, which when removed allow the technology to still compete in the open market) or "stretches and transforms" where the niche protected space becomes institutionalized into a new regime. Other important considerations discussed in the strategic niche management literature include positioning to maximize windows of opportunities, how to withdraw the protection, how to scale up niche developments, and how to establish the broader infrastructure and networks (e.g., maintenance) required to sustain the niche moving forward.

A core idea of strategic niche management, and where governments can play a key role in this approach, is that it gives niches a more equitable platform to challenge incumbent regimes. However, strategic niche management has been criticized for a number of limitations or challenges including how to get the right balance between protection and selection pressure (keep protection for too long and the niche may not be able to survive on its own), how to roll back protection measures (providing certainty to all actors), and how to accelerate development (to challenge the regime) (Heiskanen et al. 2015; Hommels et al. 2007; Kemp et al. 1998; Schot and Geels 2008).

There are examples where governments have used policies, finance, or other support mechanisms to provide protected space for niche development in the built environment, showing building regime actors what is possible within a semi-protected space while at the same time allowing for a "living laboratory" where real-world outcomes of sustainable built environment innovation can be tested and evaluated (Brown and Vergragt 2008; Geels and Raven 2006). For example, in 2004, the State Government of South Australia used a range of policies and planning requirements to establish Lochiel Park, a model green village to challenge energy and resource consumption boundaries of traditional construction and living. Lochiel Park is considered a success across a range of performance indicators (e.g., 66% reduction in energy use, 74% reduction in greenhouse gas emissions) (RenewalSA 2016). The broader impact on the building regime from this exemplar development was less clear. Berry et al. (2013) found some evidence that stakeholders directly involved in the development had gained significant experience in both design and construction of sustainable homes and had translated this to their work outside of Lochiel Park. The development was also beneficial for key policy makers through exposure to what such developments meant in practice. However, it was thought that influences on other regime actors not directly involved were limited as evidenced by only a small percentage of building regime actors engaging with the specially developed training program about the development. The tenacity of incumbent housing regimes has also been found elsewhere (Crabtree and Hes 2009; Heiskanen et al. 2015).

Sustainability transitions researchers are now asking why some niche projects are more successful than others, both in their establishment and influences on the broader regime, and how can government actors improve their role in facilitating the transition of niche practices to broader, mainstream application, through strategic niche management. This chapter explores three cases from Australia where the Federal Government or Victorian State Government[1] proactively developed policy and financial levers to accelerate development of sustainable house retrofit

[1]Australia is governed by three levels of government: federal, state, and local. The federal government is responsible for matters which affect the whole country such as defense, trade, foreign affairs, immigration, national public works, higher education, elements of healthcare, taxation, and welfare. Each state (or territory) then has its own government with their own constitutions and structure of legislature, executive, and judiciary. The state governments are responsible for matters not controlled by the Commonwealth under Section 51 of the Constitution such as state-based education, health (e.g., public hospitals), public transport, main roads, public housing, and law and order. The local government (or council) is responsible for providing services in a local area such as maintenance of local roads, rubbish collection and disposal, child care services, libraries, and town planning and local building regulations. Constitutional responsibility for local governments sits within the state and territory governments.

(residential solar photovoltaics, residential ceiling insulation) and new sustainable housing (sustainable high-density residential development). The questions this chapter seeks to address across the cases are:

- Why do certain strategic niche management projects succeed and diffuse their lessons learned to the broader regimes, while others do not?
- Given an understanding of the above question, what can government do better to improve the likelihood of successful niche developments?

7.2 Methods

This chapter explores three cases from the Australian context where the government has strategically implemented policy and financial approaches to help develop a sustainable housing niche. These cases all occurred across a similar time frame (2007–2011) and were deemed by commentators to have three different outcomes (i.e., successful, limited success, and unsuccessful). This makes the cases interesting for comparison and drawing out key learnings relating to their outcomes but also if, or why, they influenced the broader regime, within a similar (landscape) period.

Two of the cases, residential rooftop solar photovoltaics and residential ceiling insulation, were projects driven by the federal government, while the Nicholson higher-density demonstration development was facilitated by the Victorian State Government. The residential solar and ceiling insulation cases are explored through a desktop review of both schemes, including drawing upon the policy documents, public comment, and academic and consultancy reviews into the schemes. The Nicholson case presents analysis from a post-occupancy evaluation conducted by the author, and others, in 2014 (Moore and Higgins 2016; Ridley et al. 2014). Specifically, it draws upon semi-structured interviews with 14 key building industry stakeholders including developers, bankers, investors, and valuers from Melbourne, Australia, to explore what the broader influence of the development was on shaping or challenging the incumbent building regime.

7.3 Cases

7.3.1 Residential Solar Panels

7.3.1.1 What Happened?

A market-based solar photovoltaic (PV) rebate scheme was developed and administered by the Australian federal government in 2000, within a wider renewable energy policy setting (Sivaraman and Horne 2011, see also Chap. 1 by Horne). Initially the rebate offered \$5.50[2] per watt, up to a maximum of \$8250 per household. The rebate was typically claimed directly by PV installers who passed this saving onto the household (Macintosh and Wilkinson 2010). The scheme was revised numerous times, including at the change of federal government from Liberal to Labor in 2007 where the rebate amount was again increased. This triggered a rapid uptake of PV systems and unexpectedly oversubscribed the scheme leading to the scheme's termination (almost overnight) in 2009 before a weaker version was relaunched in the same year (Macintosh and Wilkinson 2011). This explosion of solar PV and similar knee-jerk policy change has occurred in other countries as well such as the USA, Spain, the UK, and Germany (Smith et al. 2014; Hess 2015).

The rebate scheme did not operate in isolation. By 2008 feed-in tariff (FiT) schemes had been introduced by each state government (as energy generation and pricing are a state responsibility) to complement the rebate scheme. While traditional electricity retail cost ~15–20 c/kWh at the time, the FiT for net energy (paid by energy providers) was up to 68 c/kWh. Due to high uptake by households and the perception that it was inflating retail energy prices across the energy network, the FiT price was reduced numerous times; as of 2016, it is 6–8 c/kWh for new solar installations. This reduction in FiT has changed the overall approach to how households use their renewable energy, switching from an incentive for households to export as much energy as possible to consuming as much as they can.

The rebate and FiT policy levers coincided with a global price drop for solar PV systems and the increasing value of the Australian dollar. This lead to a "perfect storm" whereby there were high financial incentives being offered by the federal and state governments, alongside marketplace efficiencies.

7.3.1.2 What Was the Outcome?

A review conducted in 2010 on behalf of the federal government concluded that the rebate scheme was not the most cost-effective way to reduce greenhouse gas emissions and, moreover, most of the technology was being imported from overseas (Macintosh and Wilkinson 2010). However, the report failed to consider how the solar PV industry in Australia prior to the rebate was essentially nonexistent (Zahedi 2010). The community-wide proliferation of solar PV installations means that a local solar PV (and other renewable energy technologies) industry has emerged, allowing choice of technologies, brands, suppliers, and aftermarket support. Furthermore, there are now companies who are designing and manufacturing solar PV products in Australia again. This is all evidence of the development of a broader socio-technical configuration. There is some evidence that the high uptake of residential solar PV in Australia is challenging the incumbent energy regime with several of the most polluting coal-fired power plants closing or reducing output over

[2] All references to dollars in this chapter are for Australian dollars.

the past couple of years. There is also policy discussion beginning to occur around if, or how, local sharing or renewable energy could occur (i.e., between neighbors).

There are now more than 1.57 million households with PV in Australia, or 17% of total households, for a total installed capacity of 5.4 gigawatts (APVI 2016). While the rebates and FiT have been significantly wound back in recent years, there remains strong growth for solar PV in Australia indicating that broader societal changes have taken hold (see Figs. 7.1, 7.2, and 7.3 and Chap. 1). It is possible that

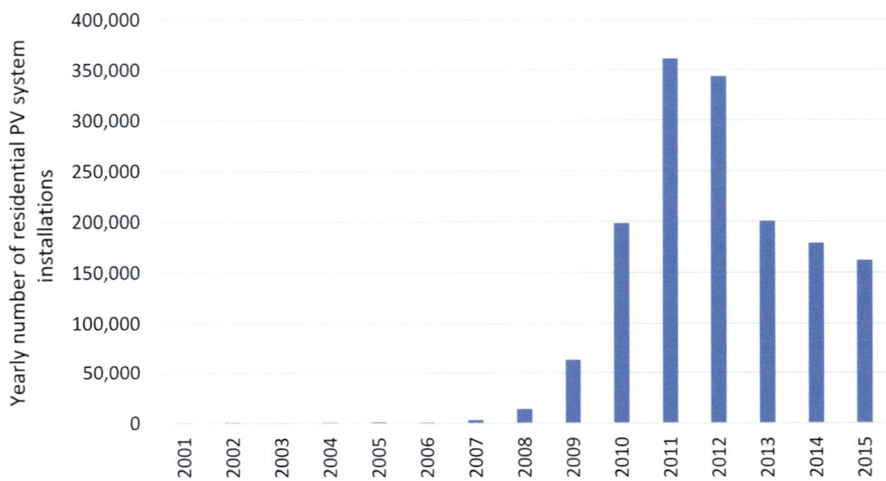

Fig. 7.1 Yearly number of residential PV systems installed in Australia since 2001 (APVI 2016)

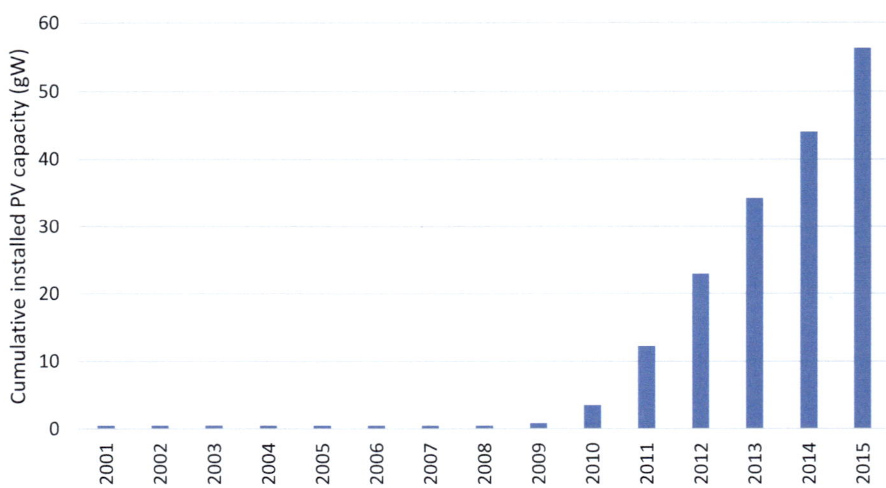

Fig. 7.2 Cumulative installed residential PV capacity (gW) in Australia since 2001 (APVI 2016)

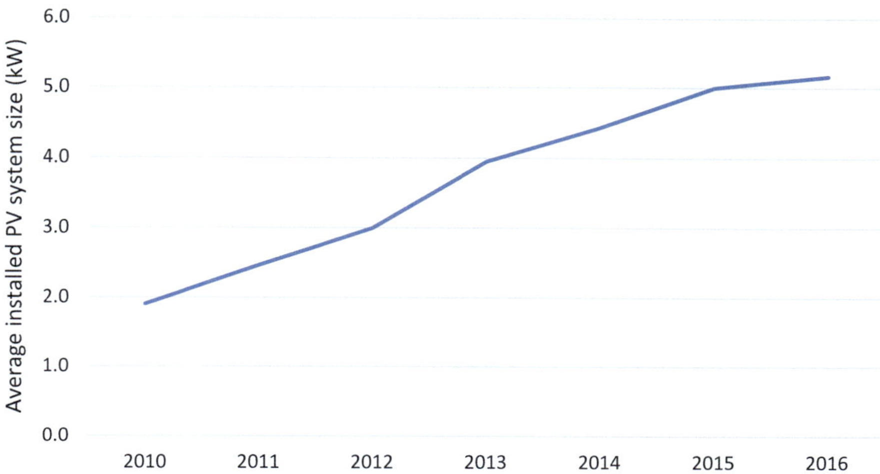

Fig. 7.3 Average installed PV system size (kW) in Australia since 2010 – note that 2016 is expected to be higher due to a lag in reporting of systems (APVI 2016)

residential PV is no longer seen as a niche development but a mainstream option for many households. The niche appears to have grown despite removal of protection.

Australian households are now well placed to build upon the uptake of solar PV with recent developments to battery storage systems to further transition away from a centralized fossil fuel energy regime. There are predictions that the price of "everyday" household battery storage systems will fall by 50% between 2014 and 2020 (Climate Council 2015). If these systems are taken up in similar numbers to the solar PV systems, there is a real opportunity to restructure the energy provision network in Australia.

The ad hoc changes to the rebate and FiT were not unexpected. These changes were going on at a time of fractured climate and general politics in Australia. However, there are signs that renewable energy is facing a slightly more certain future with several states around Australia announcing revised sustainability plans, with more ambitious goals than those set at the federal level. For example, the Australian Capital Territory has set a goal to consume 100% of energy from renewable sources by 2020, and Victoria has set a target of 40% renewable energy generation by 2025. This bodes well for a transition to a low-carbon energy future in Australia, although there is much work to get there in the face of ongoing resistance and challenges from the incumbent energy regime. This tension has only intensified after a 1-in-50-year storm caused a statewide power blackout in the State of South Australia in September 2016 where renewable energy technologies (the State generates around 40% of its energy through wind and solar energy) were unfairly blamed as the cause of the blackout (King et al. 2016).

7.3.2 *Home Insulation Program*

7.3.2.1 What Happened?

The home insulation program was developed in 2009 by the federal government primarily to stimulate the economy during the global financial crisis (ECAR Committee 2010). The program aimed to provide rebates of up to $1600 to 2.2 million owner-occupied existing homes to insulate their roof space and reduce heating and cooling energy use by up to 40%, reduce greenhouse gas emissions, and generate employment for the building industry (Hawke 2010; ECAR Committee 2010). At the time of the program's introduction, it was estimated that 2.7 million homes (40% of total stock excluding apartments) had no, or inadequate, ceiling insulation, with around 70,000 existing houses being retrofitted with sufficient ceiling insulation each year (ECAR Committee 2010).

The program started with a trial. A risk assessment of the trial from an independent consultant stated the wider rollout should be delayed by at least 3 months to address some risks with governance and implementation of the program. However, after some tweaks to the program, the federal government felt they had addressed concerns and pressed ahead with the wider roll out. From the beginning, the program was beset by a number of issues around changing standards, issues with rebate payments, rogue installers, installers with limited experience, and product safety concerns (ECAR Committee 2010; Hawke 2010). For example, the rebate amount was reduced to $1200 mid-program, the requirements for accreditation for installers were changed ad hoc, and there were continuous changes to the material installation requirements as a reactive safety measure (e.g., requiring covers over downlights).

The deaths of four contractors and over 200 house fires linked to the program (mainly from faulty wiring and installing the insulation too close to downlights) led to the abrupt discontinuation of the program within a year of starting (Dollery and Hovey 2010; Hanger 2014). At least two of the deaths were directly related to the use of the foil insulation, with a later review stating that the inclusion of this insulation material was "fundamentally flawed" considering the risk advice received before the program began (Hanger 2014, p. 3). The government was warned about the significant dangers from reflective foil laminate insulation (i.e., stapling through the aluminum, deaths of installers in New Zealand) but allowed its use in the program before banning the material only 1 week before the overall home insulation program was terminated (Hanger 2014).

7.3.2.2 What Was the Outcome?

In total, 1.2 million dwellings received insulation through the program, at a total cost to the government of $1.5 billion dollars (ECAR Committee 2010). This was a significant increase on the estimated 70,000 dwellings per year which had received ceiling insulation in years prior (ECAR Committee 2010). There is little information as to what the influence has been on installation numbers after the conclusion of the program, although it has certainly not remained at the high levels seen during the program.

It was estimated that the insulation reduced heating and cooling energy by 35% and saved households $300 in utility bills a year (BZE 2013; ICANZ 2011; Hanger 2014). There are also health benefits of improved thermal comfort for occupants, but this was not quantified in any of the program evaluations. The per house benefits were in line with the program aims, although due to the program's early termination the cumulative benefits were lower than expected.

Despite these benefits, the program is roundly seen as a failure in Australia and underwent a Royal Commission review (Hanger 2014). The program was criticized for rushing to implementation, not addressing sufficiently risk concerns identified by the independent consultant from the trial, changing numerous elements during the program, and that the program management did not have sufficient resources to service the program adequately (Hanger 2014; ECAR Committee 2010). Overall it was both the government department managing the program and the insulation industry itself, which could not handle the unprecedented level of activity, leading to problematic outcomes. Furthermore, while the main safety issues were related to the foil insulation (rather than the bulk insulation), the program has cast a negative perception across the whole insulation industry.

Despite the program's earned negative reputation, its success ought not be discarded. For example, the issue of the house fires was a key component for closing the program early. The rate of fires/100,000 installations was found to be 3.4 times less for the first 12 months after installation and 2.4 times less for periods greater than 12 months, compared to historical ceiling insulation fire rates (BZE 2013). Furthermore, the program generated more than 10,000 jobs, although it is unclear how many of these were ongoing after the programs end (Hawke 2010). In addition, there are now a much tougher range of safety standards in relation to ceiling insulation.

7.3.3 Exemplar Development

7.3.3.1 What Happened?

Federal and state governments in Australia are supporting innovation in building design and performance outcomes through approaches such as planning controls and financial support. One such example is the Nicholson: a mixed-use building with 199 apartments located 7 km from Melbourne's CBD, completed in 2011. The building site was a former tram depot owned by the Victorian State Government. With funding provided through the federal government's global financial crisis stimulus programs, state government actors saw an opportunity to push for more innovative building outcomes. The Nicholson was developed by Places

Victoria – the Victorian Government property development agency – and designed integrating the following innovative elements (Places Victoria 2015):

- Housing affordability: The inclusion of 89 affordable housing units in a "salt and pepper" approach[3] to be rented to low-income households
- Environmental sustainability: Designed to a 6-star Nationwide House Energy Rating Scheme thermal energy rating (heating and cooling load of 114 MJ/m^2.annum), which was above the 5-star minimum requirement
- Modular construction: The largest development at the time in Melbourne to use modular construction resulting in the development being completed 50% faster than traditional construction approaches

While, individually, none of the elements were particularly novel, combined, they represented significant innovation in the Melbourne housing market at that time. Places Victoria hoped the development would be a commercially replicable demonstration project and influence other developers.

7.3.3.2 What Was the Outcome?

Modular Construction

Interviews with broader building stakeholders in Melbourne found that the building was seen as an exemplar for modular construction methods. Some stakeholders were convinced that it was a legitimate approach after seeing the end product; others still had ongoing concerns around if it was cost-comparable to traditional building approaches. The developers who were interviewed had shifted to including modular elements, such as bathrooms, into their developments, providing a fit and conforming niche outcome.

A key constraint on modular construction had been around how financial institutions would fund a development which was being constructed off-site (Moore and Higgins 2016). Funds have typically been provided at certain stages of on-site development, but there was a lack of framework as to how this applied for off-site construction. The Nicholson development was recognized as helping to influence and change financial institute processes to allow funding for modular construction, removing one of the main barriers to its uptake.

Environmental Sustainability

While many stakeholders agreed that improving sustainability outcomes was the "right thing to do," most deferred back to what the market directed. The perception

[3]Where the different tenure types are located next to each other throughout the building with no discernible difference in unit design or finish quality

was that improved environmental sustainability had a negative impact on upfront affordability for consumers. However, there was evidence that some consumers were seeking out more sustainable options. This consumer market has been growing across recent years in Melbourne. Privately developed innovative and sustainable developments such as the Nightingale and the Commons are not only demonstrating improved sustainability and liveability but are challenging wider housing paradigms by including shared communal areas (e.g., laundries) and setting legal requirements so that apartments can only be sold in the future for no more than the average price rise for apartments in those suburbs (Perinotto 2015)[4]. The proof of success in these developments is the long waiting lists to buy apartments in these developments.

The impact of environmental sustainability from the Nicholson development was limited. The national minimum building requirements changed before the completion of the project, meaning that the Nicholson was a standard performing building when completed, negating the benefits of this innovative element.

Mixed Tenure

The inclusion of mixed tenure provided the most divisive discussion point among the stakeholders. Many of the stakeholders felt that Places Victoria demonstrated significant leadership and bravery in funding a mixed tenure development although most stated that it was not something they would deliver due to the perceived stigma attached. The perception was that such housing attracts a lower quality of occupant, impacting on liveability and property values. The initial sale of the Nicholson apartments would indicate that the inclusion of mixed tenure accommodation had less impact on value than first thought (Moore and Higgins 2016). This stigma was not based on any evidence but is more a longer-term cultural issue in Australia. The Nicholson was seen as an exemplar development which the building industry could learn from with regard to mixed tenure housing, although the stakeholders wanted further information about if mixing tenure was impacting on sales value and residents' lived experience before making any move toward this approach.

Broader Influence

The Nicholson won awards for its design including the 2011 Urban Development Institute of Australia Judges' Award. This afforded the project with a broader level of awareness across the building industry and provided some validation to the design approach taken. The building is widely cited throughout the building regime for its innovative elements. Uptake of the design and construction approaches and learnings from the Nicholson has been occurring to a limited extent, as evidence

[4]See also http://nightingalehousing.org/ for further information about these developments.

by the increasing use of modular construction (although primarily through modular bathrooms or kitchens rather than the whole building). This limited engagement with a demonstration building for other elements, such as environmental sustainability and mixed tenure, by the regime is not unexpected with the building regime recognized for being resistant to change (Crabtree and Hes 2009).

While it is evident that the development has played a role in influencing the building regime to some extent, more effort is required to engage regime actors earlier in the development process and communicate practical outcomes and learnings in a more systematic way to improve benefits. This ad hoc influence from an exemplar development has been found elsewhere. For example, Boyer (2015) found that the more successful influence of broader building regimes was when regime incumbents were involved with the design and development of the niche project as early as possible. However, this approach has challenges in ensuring that the regime does not constrain the innovation of the niche.

7.4 Discussion

Each of the cases demonstrated different outcomes in terms of niche development and influence on the broader regime. While commentators have concluded that the residential PV scheme was a success (due to high numbers of uptake), the ceiling insulation program a failure (due to poor governance and lives being lost), and the Nicholson exemplar development having limited success (with modular construction now more prevalent, but low influence of other innovative elements), all cases faced significant governance and implementation challenges. This evaluation of success is not based upon a consistent framework. For example, the PV example is based on numbers of uptake, while the ceiling insulation example is based upon how well it was governed (despite high numbers of uptake).

If we ask the question about which of these examples has now established itself as a viable alternative to and influencing the existing regime, we see that there is a different view about the success or failure of each niche. It is clear from the evidence of installation numbers that residential solar PV has had a rapid uptake by Australian households. While rebates and FiT have been substantially wound back, there is still a steady uptake from additional households. This, combined with increasing renewable energy targets by state governments and the closure of several coal-based power plants in recent years, indicates that residential solar PV is a part of the push by renewable energy technologies as a viable alternative to the existing energy regime. This is not to say there are no challenges ahead including how to include solar PV in higher-density housing (e.g., apartment buildings), how the increasing two-way flow of small energy providers (i.e., each household system) impacts on the broader energy grid, how to allow local sharing of renewable energy (e.g., sharing with a neighbor directly), and the impact of battery storage. The residential PV transition is well placed to make the next big leap with the inclusion of affordable battery storage. If there is a similar trajectory of uptake that was seen with residential solar PV panels, there is a real opportunity that the existing provision of energy in Australia will be disrupted. Traditional energy providers may have recognized that their business models are under threat as more of them are themselves selling solar PV to their customers.

While commentators deemed that the ceiling insulation program a governance failure, in terms of actual outcomes, there were successes in the form of rapid installation uptake. The program resulted in more than two million homes receiving ceiling insulation which has significant thermal and economic benefits for households and the broader environment. While there is a lingering stigma attached to ceiling insulation from the program, it is still recognized as one of the most cost-efficient ways to reduce heating and cooling requirements in dwellings and is considered "normal" within the housing regime for both new and existing housing.

Conversely, based upon low numbers of uptake of the innovative elements of the Nicholson development across the building sector, it might be said to be a failure. However, the incumbent building regime clearly saw it as an exemplar building they could learn from, even if this has not yet resulted in changes to their business-as-usual approach. Modular construction is now more prevalent across the building industry, although primarily for bathrooms and kitchens rather than whole buildings. Environmental sustainability is also gaining traction; however, this is mainly from smaller, boutique developers. Mixed tenure was one element which has not seen uptake because of the niche development. The Nicholson may be more of a "stepping stone" niche, whereby it now allows others to build upon what it achieved.

In each case, the federal or state government played a critical role in shielding and nurturing the niche allowing for a more level playing field. Although not consciously doing this through a strategic niche management approach, the government actors were developing protected spaces for these niches to flourish. This was primarily through financial assistance (e.g., rebates, FiT, provision of land for development) but, in the case of the residential solar PV, also in a broader development of supporting policies such as wider renewable energy and climate change targets. For the residential solar PV and Nicholson cases, the niche outcomes aimed to be fit and conform, so that when the financial and other support was removed, the niche would continue to develop and challenge the incumbent regime. The ceiling insulation program aimed to be a stretch and fit approach, where ceiling insulation was to be institutionalized into the regime, with ceiling insulation now considered a standard requirement within existing housing. However, in the case of the residential PV scheme and the ceiling insulation program, the governance of both was poorly planned in terms of matching the program to the speed of uptake and how to handle the administration of both and was characterized by frequent knee-jerk policy changes. This last point gave no confidence to both the industry and consumers.

While Heiskanen et al. (2015) caution against the tendency to question the value of demonstrations or niche developments in terms of short-term success, there are some clear outcomes and lessons already evident from the cases which can act as stepping stones toward more sustainable outcomes. The case studies find that when strategic niche management actions align with socio-technical landscape dynamics, then public sector actors have a greater chance at stimulating regime transformation. Inversely, when the public sector fails to align its actions with landscape forces, its experiments are seen as failures or remain unseen.

For example, the cases demonstrate the critical element of timing for governments who wish to play an active role in promoting niche developments. In the case of the solar panels, the timing was perfect, not only with landscape changes (global photovoltaic price drop, increase in Australian dollar) but also with it coinciding with two key federal and state policies which provided further financial assistance to household. However, with the Nicholson development, the sustainability features fell flat as by the time the building was complete the performance was already the regulated minimum standard across the industry, and therefore it had limited broader benefits. As for the ceiling insulation, the federal government rushed to meet the perceived opportunity (i.e., providing work during the global financial crisis) which ultimately led to the project's perceived failure. However, the success or otherwise of these cases has been assessed within a relatively short period from the introduction of the initiatives. It may be that assessment over a longer period (e.g., 20 years) will result in different outcomes, with potentially more favorable outcomes for the Nicholson development and the ceiling insulation scheme as the narrative becomes more about the ongoing longer-term benefits rather than upfront governance challenges.

The importance of windows of opportunities (or transitions points) has been identified by others in the literature (Geels and Schot 2007; Nill and Kemp 2009). It is a complex task for government actors to be able to identify potential windows of opportunity which align not only with landscape opportunities but also with niche alternatives being ready to grow. However, researchers such as Loorbach (2014) argue while precise futures are unpredictable, governments have a responsibility to align experimental outcomes with broader, more predictable trends. The Victorian Government, for example, failed to align their niche project with changing regulatory standards, ultimately rendering the result of their building experiment moot.

Once a window of opportunity is identified, the question turns to how to nurture and shield niches. This is about providing the right amount of support to allow development, but not so much that it undermines the ability of the niche to branch out on its own at some point (Coenen et al. 2010). Too much support can also lead to too quick uptake which, as demonstrated through the solar and ceiling insulation examples, leads to a range of governance and implementation problems, such as demand greater than the resources available. Nill and Kemp (2009) present an example where the Dutch wind energy sector was provided with too much support which resulted in manufactures of wind turbines sticking with suboptimal design.

Shielding and nurturing a niche do not automatically translate to a new outcome for the regime which is likely to challenge and resist changes or where poorly planned policy outcomes emerge (Kemp et al. 1998). As Smith et al. (2014, p. 128) state:

> ... demonstration programmes become political exercises as well as technology assessment measures. Demonstration programmes provide opportunities for the build-up of advocacy, widening participation, introducing new framings and narratives, re-assessment of interests as well as the technology.

But even in these cases where there were challenges, many transitions researchers argue that understanding these as sites of learning about the niche and its engagement with the broader regime is just as important as understanding successful transitions (Heiskanen et al. 2015). In the case of the ceiling insulation scheme, the program underwent a Royal Commission review, covering all successes and failures of the program in detail. With the review finding that many of the failures of the program were identified and raised by various actors prior to the program rollout, the question is will the learnings inform future program development and roll out efforts?

An ongoing challenge for strategic niche management approaches is how to remove the protection to allow the niche to challenge the regime on its own merit. The residential solar example from Australia, and as has occurred in other countries, is one of the ad hoc policy developments and changes. Almost overnight support was changed on several occasions which demonstrated there was no clear pathway to winding back support when the scheme reached certain levels. The abrupt changes by the government did not provide certainty of direction for the solar industry or consumers. The fact that the residential PV sector continues to grow despite the confusing politics is testament to how strong the niche has become and how opportune the window of opportunity was. At the height of the government support, households were being sold on the dream of eliminating energy bills all together. Whether that happened requires further research, but consumers went on and continue to install the technology regardless. On the other hand, the lessons from the Nicholson seem not to have been communicated in a compelling or systematic way. While the government commissioned a post-occupancy evaluation (Ridley et al. 2014), it does not seem like wider dissemination of the project occurred across the building regime. In their discussion, Berry et al. (2013, p. 8) identify one of the key challenges of the Lochiel Park demonstration project:

> Greater effort could have been made to achieve widespread learnings across the local building sector, and deeper level learnings to a larger number of industry and policy professions, and thus increase the influence of the single niche event on the local housing sector.

The examples demonstrate that governments are key actors in guiding sustainability transitions. If government actors want to improve the likelihood of successful niche developments, they should work to identify windows of opportunity at both the landscape level and at the niche/regime levels. Furthermore, developing policy or programs that have sufficient resources to administer them is critical.

7.5 Conclusions

This chapter explored three different cases from Australia of strategic niche management. It is evident that some of the niches were more successful, in terms of their establishment as a viable alternative to the existing regime and their influence over that regime than others. The case studies show that when strategic niche management actions align with socio-technical landscape dynamics, then public sector actors have a greater chance at stimulating regime transformation. Inversely, when the public sector fails to align its actions with landscape forces, its experiments are seen as failures or remain unseen in the short term. Within this, context windows of opportunity are critical to improving the likelihood of success for niches and to ensure they can influence the broader regimes.

Overall though, government actors, both in Australia and internationally, must play a role in shaping and directing urban sustainability transitions. Governments have significant levers at their disposal to affect outcomes and deliver more even playing fields for niche challengers through strategic niche management if they can align their actions with socio-technical landscape dynamics and provide sufficient protection (e.g., policy development, rebates). The challenge transitions researchers, urban planners, and policy makers face is we must all work toward a better understanding of how to ensure successful niche developments which can challenge existing regimes if a transition to a more sustainable future is to occur.

Acknowledgment The discussion about the Nicholson development is informed by research conducted by the author and David Higgins (Moore and Higgins 2016) which was undertaken for a project funded by Places Victoria. The author also acknowledges the wider project team involved in evaluating the Nicholson development including Ian Ridley, Jin Woo, Karishma Kashyap, Megan Nethercote, and Ralph Horne.

References

APVI (2016) Australian PV market since April 2001. Australian PV Institute. Available at: http://pv-map.apvi.org.au/analyses. Accessed 27 June 2016

Berry S, Davidson K, Saman W (2013) The impact of niche green developments in transforming the building sector: the case study of Lochiel Park. Energy Policy 62:646–655. Available at: http://linkinghub.elsevier.com/retrieve/pii/S0301421513007106. Accessed 20 Feb 2014

Boyer RHW (2015) Grassroots innovation for urban sustainability: comparing the diffusion pathways of three ecovillage projects. Environ Plan A 47(2):320–337. Available at: http://epn.sagepub.com/lookup/doi/10.1068/a140250p. Accessed 27 June 2016

Brown HS, Vergragt PJ (2008) Bounded socio-technical experiments as agents of systemic change: the case of a zero-energy residential building. Technol Forecast Soc Chang 75:107–130

BZE (2013) Zero Carbon Australia Buildings Plan, Melbourne Available at: http://media.bze.org.au/bp/bze_buildings_plan.pdf. Accessed 25 June 2016

Caniëls M, Romijn H (2006) Strategic niche management as an operational tool for sustainable innovation: guidelines for practice. Eindhoven Center for Innovation Studies, Heerlen

Climate Council (2015) Powerful potential: battery storage for renewable energy and electric cars, Sydney Available at: http://www.climatecouncil.org.au/uploads/b5719aa238223c1b2acb126f734fc1fe.pdf

Coenen L, Raven R, Verbong G 2010 Local niche experimentation in energy transitions: a theoretical and empirical exploration of proximity advantages and disadvantages. Technol Soc 32(4):295–302. Available at: http://www.sciencedirect.com/science/article/pii/S0160791X10000734. Accessed 28 May 2014

Crabtree L, Hes D (2009) Sustainability uptake in housing in metropolitan Australia: an institutional problem, not a technological one. Hous Stud 24(907688161):203–224

Dollery B, Hovey M (2010) Australian Federal Government Failure: the rise and fall of the home insulation program. Econ Pap J Appl Econ Policy 29(3):342–352. Available at: http://dx.doi.org/10.1111/j.1759-3441.2010.00079.x

ECAR Committee (2010). Environment, Communications and the Arts References Committee. Energy efficient homes package (ceiling insulation), Canberra

Geels FW (2011) The multi-level perspective on sustainability transitions: responses to seven criticisms. Environ Innov Soc Trans 1(1):24–40. Available at: http://www.sciencedirect.com/science/article/pii/S2210422411000050. Accessed 23 May 2014

Geels F, Raven R (2006) Non-linearity and expectations in niche-development trajectories: ups and downs in Dutch biogas development (1973–2003). Technology Analysis & Strategic Management. Available at: http://www.tandfonline.com/doi/abs/10.1080/09537320600777143?journalCode=ctas20. Accessed 14 Feb 2016

Geels F, Schot J (2007) Typology of sociotechnical transition pathways. Res Policy 36(3):399–417

Hanger I (2014) Report of the Royal Commission into the home insulation program, Canberra. Available at: http://www.homeinsulationroyalcommission.gov.au/documentation/documents/reportoftheroyalcommissionintothehomeinsulationprogram.pdf

Hawke A (2010) Review of the administration of the home insulation program. On behalf of the Department of Climate Change on Energy Efficiency, Canberra. Available at: http://www.climatechange.gov.au/~/media/publications/energy-efficiency/Home-Insulation-Hawke-Report.ashx

Heiskanen E, Nissilä H, Lovio R (2015) Demonstration buildings as protected spaces for clean energy solutions – the case of solar building integration in Finland. J Clean Prod 109:347–356

Hess DJ (2015) The politics of niche-regime conflicts: distributed solar energy in the United States. Environ Innov Soc Trans 19:42–50. Available at: http://www.sciencedirect.com/science/article/pii/S2210422415300174. Accessed 18 Mar 2016

Hommels A, Peters P, Bijker WE (2007) Reply to Geels and Schot. Res Policy 36:1102–1103

ICANZ (2011) The value of ceiling insulation. Impacts of retrofitting ceiling insulation to residential dwellings in Australia, Melbourne. Available at: http://icanz.org.au/wp-content/uploads/2013/04/ICANZ-CeilingInsulationReport-V04.pdf

Kemp R, Schot J, Hoogma R (1998) Regime shifts to sustainability through processes of niche formation: The approach of strategic niche management http://dx.doi.org/10.1080/09537329808524310

King A et al (2016) What caused South Australia's state-wide blackout? The conversation. Available at: http://theconversation.com/what-caused-south-australias-state-wide-blackout-66268. Accessed 13 Oct 2016

Living Building Challenge (2015). zHome. Available at: http://living-future.org/case-study/zhome. Accessed 24 Nov 2015

Loorbach D (2014) To transition! governance panarchy in the new transformation. Erasmum University, Rotterdamn

Macintosh A, Wilkinson D (2010) The Australian Government's solar PV rebate program. An evaluation of its cost-effectiveness and fairness. ACT: The Australian Insitiue

Macintosh A, Wilkinson D (2011) Searching for public benefits in solar subsidies: a case study on the Australian government's residential photovoltaic rebate program. Energy Policy 39(6):3199–3209. Available at: http://www.sciencedirect.com/science/article/B6V2W-52HHYYF-2/2/29b65fcf8de216e2d15f62956d2c0799

Moore T (2014) Modelling the through-life costs and benefits of detached zero (net) energy housing in Melbourne, Australia. Energ Buildings 70(0):463–471. Available at: http://www.sciencedirect.com/science/article/pii/S0378778813008013

Moore T, Higgins D (2016) Influencing urban development through government demonstration projects. Cities 56:9–15. Available at: http://dx.doi.org/10.1016/j.cities.2016.02.010

Nill J, Kemp R (2009) Evolutionary approaches for sustainable innovation policies: from niche to paradigm? Res Policy 38(4):668–680

Perinotto T (2015) Radical apartments: after the commons, the Nightingale keeps ruffling feathers. The fifth estate. Available at: http://www.thefifthestate.com.au/business/innovators-fringe-elements/radical-apartments-after-the-commons-the-nightingale-keeps-ruffling-feathers/72333. Accessed 24 June 2016

Places Victoria (2015) The Nicholson. Available at: http://www.places.vic.gov.au/precincts-and-development/the-nicholson. Accessed 25 Nov 2015

RenewalSA (2016) Lochiel Park. Available at: https://renewalsa.sa.gov.au/projects/lochiel-park/. Accessed 21 June 2016

Ridley I et al (2014) An evaluation of the Nicholson development. RMIT University, Melbourne

Rotmans J, Kemp R, Van Asselt M (2001) More evolution than revolution: transition management in public policy. Foresight 3(1):15–31

Schot J, Geels FW (2008) Strategic niche management and sustainable innovation journeys: theory, findings, research agenda, and policy (907688161). Available at: http://dx.doi.org/10.1080/09537320802292651

Sivaraman D, Horne R (2011) Regulatory potential for increasing small scale grid connected photovoltaic (PV) deployment in Australia. Energy Policy 39(2):586–595. Available at: http://www.sciencedirect.com/science/article/B6V2W-51F819V-2/2/dc4215761d8a7d9d001dea00dc9d37d0

Smith A, Raven R (2012) What is protective space? Reconsidering niches in transitions to sustainability. Res Policy 41(6):1025–1036

Smith A et al (2014) Spaces for sustainable innovation: solar photovoltaic electricity in the UK. Technol Forecast Soc Chang 81:115–130

Urban Developer (2016) Inside Mirvac &CSR's new modular construction prototype. TheUrbanDeveloper.com. Available at: https://www.theurbandeveloper.com/inside-mirvac-csrs-new-modular-construction-prototype/?omhide=true&utm_medium=email&utm_campaign=MON VIC&utm_content=MON VIC+CID_811085678a323ee36e872f2ec50fe576&utm_source=Campaign Monitor&utm_term=Inside Mirvac CSRs Ne. Accessed 24 June 2016

Zahedi A (2010) Australian renewable energy progress. Renew Sust Energ Rev 14(8):2208–2213. Available at: http://www.sciencedirect.com/science/article/pii/S1364032110000882

Part III
Specific Approaches to Urban Transitions

Part III presents different sectorial perspectives on the application of transitions in the urban through three chapters. Chapter 8 discusses urban water management in the context of transitioning to Water sensitive cities through a case study of Melbourne, Australia. Chapter 9 explores alternative approaches to housing population growth via sustainable redevelopment of greyfield precincts by applying a transitions management approach and the use of transitions arenas. Chapter 10 concludes this section of the book by exploring the Transition Towns movement and the role such an approach could have on broader urban transitions, illustrating outcomes through a visioning exercise.

Chapter 8
A Framework to Guide Transitions to Water Sensitive Cities

Rebekah R. Brown, Briony C. Rogers, and Lara Werbeloff

Abstract This chapter explores the transition challenges and opportunities facing urban water sectors globally, as pressures from urbanisation, climate change and ecological degradation drive new approaches to urban water management. Framed around the vision of a future Water Sensitive City and drawing on empirical evidence from a case study of storm-water quality management in Melbourne, Australia, this chapter presents a framework for benchmarking a city's progress in its urban water transition. It provides a nuanced understanding of how these complex change processes unfold and identifies the enabling conditions that can be used to help steer change in urban water systems towards the envisioned water sensitive city.

Keywords Benchmarking framework • Institutional change • Stormwater quality management • Urban water transitions • Water sensitive cities

8.1 An Emerging Vision for Urban Water Systems

Water management in the twenty-first century cities has become increasingly challenging. For the first time ever, urban populations exceed those in rural areas, and the pressure on water systems in cities is growing accordingly. Traditional water management approaches, typically based on principles of predictability and control, are increasingly ill-equipped to cope with today's complex and interrelated

R.R. Brown (✉) • L. Werbeloff
Monash Sustainable Development Institute, Monash University, Melbourne, VIC, Australia

Cooperative Research Centre for Water Sensitive Cities, Monash University, Melbourne, VIC, Australia
e-mail: rebekah.brown@monash.edu; lara.werbeloff@monash.edu

B.C. Rogers
Cooperative Research Centre for Water Sensitive Cities, Monash University, Melbourne, VIC, Australia

School of Social Sciences, Monash University, Melbourne, VIC, Australia
e-mail: briony.rogers@monash.edu

T. Moore et al. (eds.), *Urban Sustainability Transitions*, Theory and Practice of Urban Sustainability Transitions, DOI 10.1007/978-981-10-4792-3_8

challenges including climate change, urbanisation, population growth, degraded waterways and environments, resource constraints and the liveability aspirations of communities.

With greater frequency and severity of extreme weather events such as storms, floods and droughts resulting from climate change, the flow of water through and around cities is becoming increasingly unpredictable. With this uncertainty and variability in the conditions that water systems need to be able to cope with, they must be managed adaptively. Further, systems need to be planned and designed to deliver multiple benefits including secure water supplies and flood risk management, as well as responding to issues such as the urban heat island effect and deteriorating ecological conditions of urban waterways and other urban ecosystems.

De Haan et al. (2015), in introducing a special issue on urban water transitions in the Environmental Innovation and Societal Transitions journal, provide an overview of new visions and solutions that have entered the discourse on water management globally in response to these challenges. Collectively these new approaches represent an emerging water sector paradigm that is underpinned by principles of integrated water management, hybrid systems of centralised and decentralised technologies, green infrastructures that deliver a range of ecosystem services, urban planning and design approaches that are responsive to changing water conditions, and communities that value water and are actively engaged in its management.

In the Australian context, and increasingly internationally, this new paradigm is represented by the vision of the Water Sensitive City (Wong and Brown 2009; Ferguson et al. 2013a). The vision encompasses holistic management of the integrated water cycle to protect and enhance the health of receiving waterways, mitigate flood risk and create public spaces that harvest, clean and recycle water. It uses water management as a means of facilitating better liveability outcomes more broadly and recognises that a water sensitive approach to urban development and regeneration processes can help deliver on a range of objectives critical to the liveability of a city, such as biodiversity, public green space, healthy waterways, connected communities and cultural significance. Ultimately, a water sensitive approach is underpinned by a recognition that water plays a key role in the creation of connected, vibrant and liveable communities.

Achieving the Water Sensitive City vision will require substantial shifts in the approach and attitudes among community, government and business sectors. Ferguson et al. (2013b) describe the range of cultural-cognitive, normative and regulative dimensions that would need to be addressed to establish an enabling institutional context for supporting these system-wide shifts. For example, creating new understanding of our inability to predict and control the environment, generating knowledge and evidence to support technological innovations, establishing new water policy goals associated with urban liveability outcomes, providing political leadership to support implementation of the new paradigm, encouraging communities to be vocal about their water-related aspirations and expectations, instituting water governance arrangements that support collaboration, and designing markets that are receptive to innovative modes of water service provision.

The Water Sensitive City vision is part of a broader transition towards sustainable cities. Similar transformations are currently unfolding in other domains including energy (See Chaps. 1, 7 and 11), urban development (see Chap. 9) and transport (see Chap. 13). Transformations towards sustainability across these various sectors will ultimately need to integrate and co-evolve with each other in order to deliver urban transitions more broadly.

There is a growing body of evidence on the socio-institutional barriers inhibiting progress towards these types of enabling conditions in the water sector (e.g. Brown and Farrelly 2009). In response, urban water strategists, policymakers and practitioners are now asking how they can actively work towards overcoming these barriers to establish an institutional context that is conducive to facilitating the required system changes on the ground.

This chapter aims to provide guidance on this question, drawing on empirical evidence from a case study of transition in Melbourne's stormwater quality management system to develop a framework for benchmarking and navigating a city's transition to a Water Sensitive City.

8.2 Urban Water Transitions

The Water Sensitive City vision represents a significant change from the current water management approach in Australian cities, meaning that transitions are needed. Ideas and concepts from transitions studies offer useful insights into the dynamics of transition processes within complex systems and provide theoretical guidance on how such change processes may be steered. For the purpose of this chapter, transitions are understood to be a multidimensional transformative change process whereby a system shifts towards more sustainable modes of production and consumption (Markard et al. 2012). They typically take place over 25–50 years and are characterised by complementary changes across and within a number of domains (including technological, economic, institutional, behavioural and cultural), all of which operate synergistically to reinforce and drive the transition (Rotmans et al. 2001).

For large sectoral systems such as urban water, transitions rely on overcoming institutional inertia. The phenomenon of institutional inertia and lock-in is well documented, whereby suboptimal technologies and approaches are maintained because of the sheer enormity of the task to change them (Pahl-Wostl et al. 2009; Walker 2000). Sunk costs associated with prior infrastructures, institutions and entrenched routines and practices also operate as significant barriers to the adoption of better alternatives (Pierson 2004). Even where there is impetus and appetite for change, such change will typically unfold in an incremental, path-dependant way. Deeply embedded rules and behaviours also help to keep the current framework firmly anchored (Pierson 2004). A combination of technological lock-in, institutional inertia and the challenge of reorienting professional and organisational capacity towards a new approach all serve as significant barriers to sector-wide transitions. As such, more radical sustainability approaches that may not be consistent with the established architecture are immediately less favourable, with cities instead tending to continually optimise inherently unsustainable approaches.

Realisation of such transitional change in urban water sectors is no easy task. Like the systems for provision of all essential services, the current urban water management framework is the result of complex interactions across a range of dimensions, including regulatory, institutional, infrastructure, markets, behaviours, user practices and technical expertise, which have all co-evolved over a long time and operate in a mutually reinforcing way to embed the status quo.

Research into sector-wide transition processes indicates that realisation of change on the ground requires mutually reinforcing change across infrastructures, institutions and practices (De Haan and Rotmans 2011; Smith et al. 2005; Rotmans et al. 2001). Further, a focus on technical innovation is not enough and that the social and institutional dynamics that underpin any city's transition attempt is key when trying to move entrenched water management systems into new directions (Rogers et al. 2015). Effectively managing these various components of the change process therefore relies on concerted effort, requiring ongoing commitment, monitoring and investment to steer change in desirable directions.

To consider transition processes in the context of urban water management more specifically, the Urban Water Transitions Framework (Brown et al. 2009) is a useful guide. The framework identifies the evolving sociopolitical drivers and service delivery functions as six distinct developmental 'states' that cities can be seen to move through, or are expected to move through, in response to society's increasing liveability, sustainability and resilience aspirations for urban water management (Fig. 8.1). These city states form an embedded continuum, culminating in the Water Sensitive City.

A critical step on the path towards a Water Sensitive City is to overcome the barriers in moving beyond a *Drained City*. As reflected in Fig. 8.1, moving to the right side of the continuum is particularly challenging, as it requires a significant reorientation of existing infrastructures, institutions and approaches to water management. The water servicing needs of the three left side city states have traditionally been met through large-scale, centralised infrastructure. However, the more complex and interrelated needs of the three right city states require a shift to an interdisciplinary approach that provides more flexible and integrated infrastructures and institutions at both centralised and decentralised scales.

In the *Waterways City*, the social, environmental and aesthetic values of clean waterways are a key focus, with consequent investment in improved stormwater management. This is particularly challenging, as stormwater pollution cannot be managed successfully through existing centralised technology but requires a holistic approach, including both source control and distributed systems in combination with centralised infrastructure.

In the *Water Cycle City*, water and other resources (such as energy and nutrients) are actively conserved and regenerated; supplies from diverse sources such as storm water, greywater and recycled wastewater are put to their most appropriate uses. Across social, economic and environmental contexts, sustainability is widely embraced, and the former hydro-social contract, in which government was expected

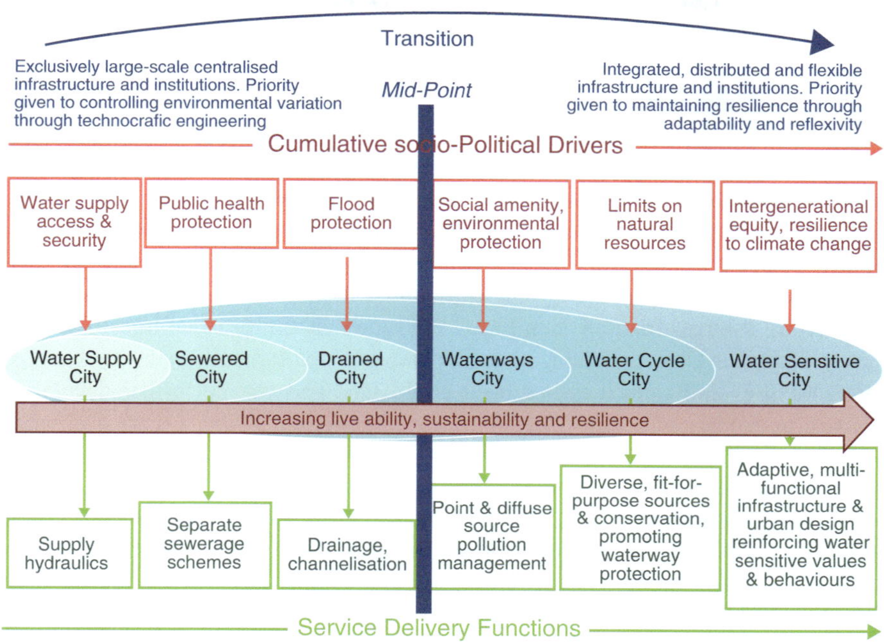

Fig. 8.1 Urban water transitions framework (Adapted from Brown et al. 2009; De Haan et al. 2015)

to deliver risk-free water supply services, is replaced with co-management arrangements between government, business and community. Such progress is hampered where governments remain resistant to change. In Australia, for example, recent extended drought conditions led to the return of rhetoric for greater control and expansion of centralised systems, rather than to the implementation of a flexible, shared approach.

Finally, the futurist state of the *Water Sensitive City* will be very different from conventional urban water management approaches. It is envisaged that communities will be driven by the aim of actively protecting natural resources and ecological integrity and providing resilience to climate change for future generations. Infrastructure, technology and urban design will be flexible, recognising the link between society, technology and the urban form. All parts of the water sector will be committed to the sustainable management of a city's water resources and hydro-social contracts will be adaptive and evolving. This state represents the ultimate vision outlined by the Urban Water Management Transitions Framework, an endpoint to be attained through overcoming current challenges. The Water Sensitive City vision therefore represents a fundamentally different approach to water management in our cities, requiring new infrastructural, institutional and organisational capacities.

8.3 Melbourne's Transition to a Waterways City

As Fig. 8.1 highlights, the realisation of the Water Sensitive City vision relies on transitional change, involving the development of responses to the increasingly complex sociopolitical drivers on the right side of the framework. Shifts from one city state to another, while remaining in the left-hand side of the framework (e.g. from the Sewered to Drained City), represent incremental developments or evolutionary steps. Similarly, shifts between states on the right side are also evolutionary in nature. However, the shift across the midpoint of the Urban Water Transitions Framework is a significant milestone in the broader transition to a Water Sensitive City, indicating that a foundation for radically new institutions, infrastructures and capacities has been established and can be further developed to assist realisation of a Water Sensitive City.

In order to explore the transition dynamics of urban water systems in more detail, we now present the story of this transition for Melbourne, based on a longitudinal case study by Brown et al. (2013). This case study is the first evidence-based investigation of how a city can transition towards more water sensitive practices; understanding the enabling social and institutional dynamics can provide valuable insights into how other cities may be able to move entrenched water management systems in novel directions.

The metropolitan region of Melbourne, Australia, has achieved significant milestones in its transition towards sustainable urban water management. Over the past 50 years, the city changed its stormwater management from a traditional drainage system that releases untreated stormwater into rivers and the ocean to one that retains stormwater at source to support multiple objectives, such as creating local ecological landscapes, improved amenity and flood mitigation. First commencing in the 1960s, this decades-long shift has placed Melbourne ahead of other Australian cities through the provision of a city-wide market-offsets scheme, as well as through a state-government regulatory mandate for sustainable stormwater management that is applied to all new developments across Melbourne. The city now boasts a large number of stormwater treatment development projects across metropolitan Melbourne and is actively engaging municipalities as well as private landholders in the process. The guidance provided by a number of champions from across the community, government, as well as the private and research sectors has been instrumental in replacing outdated perspectives on water governance with new approaches that meet the changing needs of the twenty-first century.

While the transition in urban stormwater management is only one component of a broader WSC transition, it is an important step on the transition journey. The diffuse nature of stormwater pollution, and consequent need for solutions to be integrated into the broader urban landscape, requires a significant shift in the dominant infrastructures, institutions and practices that typically characterise city states on the left-hand side of the continuum. The success of this shift over this transition milestone and into the Waterways City is indicated by a widespread new practice in Water Sensitive Urban Design, an extensive policy and regulatory framework and a new network of actors involved in both policy and practice spheres. It is therefore a case worthy of examination in order to better understand the dynamics of how such revolutionary shifts unfold in practice.

From its current position in the Waterways City, Melbourne has a strong foundation to transition more swiftly towards the WSC vision. As noted above, the shift from the left to the right-hand side of the continuum is a significant transition milestone, and subsequent shifts to the other city states require only evolutionary or incremental changes that build on the foundation already established. As a result, there is the potential for Melbourne to move rapidly towards the WSC vision, although the pace of change will of course be influenced by a range of factors including political will and industry and government investment.

Evidence from Melbourne's transition from a Drained City to a Waterways City, based on extensive qualitative and quantitative data spanning over five decades, reveals a typology of six distinct phases through which the city's transition towards more sustainable stormwater management was seen to progress (Fig. 8.2). These phases were developed through a chronological analysis of Melbourne's transition, which was derived from an extensive process of data collection, comprising oral history (n = 28), individual (n = 24) and group (n = 91) interviews (Brown et al. 2013). The analysis went through a validation process with over 50 interviewees representing all stakeholders related to urban stormwater quality management in Melbourne to identify gaps in the historical account, refine the transition phases and domains of change identified (Brown et al. 2013).

Taken together, these six phases chart the initial emergence of a sustainability issue through to the eventual embedding of new sustainable water management practices as part of a business-as-usual approach. Each phase was characterised

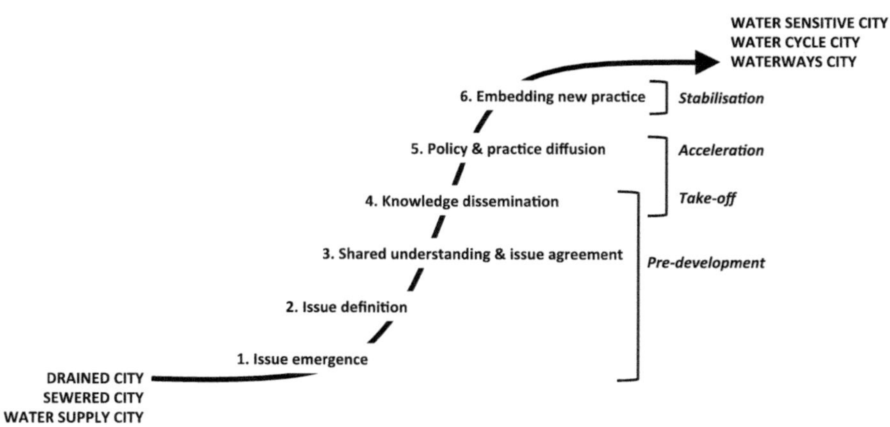

Fig. 8.2 Six phases in the transition towards water sensitivity (Adapted from Brown et al. 2013)

by different social and institutional dynamics and presented unique challenges and opportunities for strategic intervention to improve water management policy and practice.

In the *issue emergence* phase, a particular problem is identified (e.g. poor waterway health), and in the *issue definition* phase, a cause of that problem is identified (e.g. stormwater pollution). The *shared understanding and issue agreement* phase is characterised by a common understanding of, and agreement on, the problem, its causes and its repercussions. Solutions are not yet agreed on, but the need for action is acknowledged. From this point, the *knowledge dissemination* and *policy and practice diffusion* phases are marked by greater agreement on the appropriate solutions among a broad cross section of stakeholders. The final transition phase involves *embedding the new practice* as mainstream.

The transition dynamics within each phase during Melbourne's transition to a Waterways City are now presented, highlighting the key institutional developments that worked to embed new stormwater quality management practices. Each phase is characterised by dominant narratives that represent the way a particular practice was described or talked about within the sector. The dominant narratives are seen to have evolved as the transition unfolded, with an *advocating narrative* (supportive of the new practices and promoting its uptake) increasing its power over a *contesting narrative* (challenging the new practices or suggesting they are unnecessary or inappropriate).

8.3.1 Phase 1: Issue Emergence (Mid-1960s–1989)

Reflecting the emergence and growth of the environmental movement, the public (and the media) increasingly began to question the current approaches to waterway management in the context of environmental protection. The growing social capital around healthy waterways began to exert pressure on the government to reduce waterway pollution and was given added impetus by a suite of scientific studies that confirmed the adverse impact of stormwater pollution. Once principles of environment protection became formally enshrined in law, the sociopolitical capital around waterway health provided the foundation for individual actors to create a niche for better water management within the traditional water management regime.

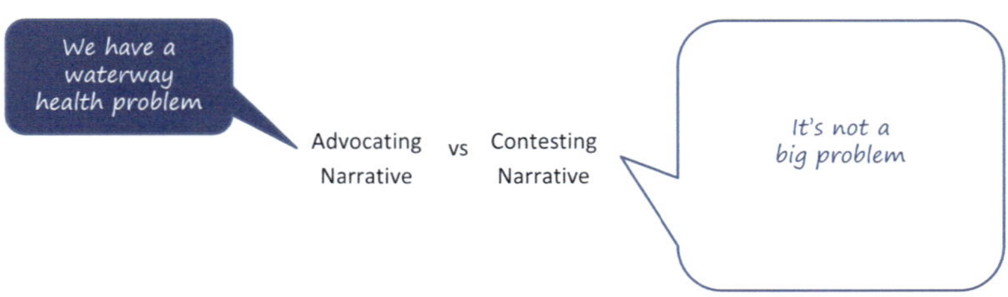

8.3.2 Phase 2: Issue Definition (1990–1995)

Sustainable urban water management took 5 years to fully emerge as an alternative vision for Melbourne's water sector. A common vision of better-managed waterways initially led to the formation of an informal network of people from backgrounds such as private engineering, academia and state and local government who had an interest in water. This group of champions emerged organically, driven by a common passion for the environment and healthy waterways, and soon found itself supported by the formation of two national Cooperative Research Centres (CRCs) with complementary foci on freshwater ecology and catchment hydrology. These CRCs fulfilled important roles as bridging organisations. Both had a strong presence in Melbourne and helped to increase the profile of better water management in this city. The CRCs produced reliable scientific information on the impact of stormwater pollution on receiving waterways and initiated the development of new technologies to deal with water quality issues. Importantly, they also led to the formation of strong collaborative relationships between Melbourne Water (the water utility responsible for the major drainage system and river health) and local universities that continue today.

8.3.3 Phase 3: Shared Understanding and Issue Agreement (1996–1999)

With the issue of stormwater pollution firmly established by this phase, the network of science and industry collaborators expanded to include planners, land developers and local government representatives. During this period, the call for better stormwater management was strengthened by the realisation that the nitrogen content of stormwater run-off was polluting Port Phillip Bay. This led to the establishment of a formal stormwater policy committee, which developed best practice guidelines along with policy-linked stormwater quality run-off targets.

Another important development during this formative period was that alternative water treatment technologies were being put to the test. Local champions secured national funding to build a number of demonstration water treatment wetlands, which reassured industry of the value of this alternative technology. The technology was also tested at the larger scale of a new residential estate on the outskirts of

Melbourne (Lynbrook Estate). This project was sponsored by Melbourne Water, which agreed to be the sole risk bearer of this trial. Collectively, these highly visible projects helped to crystallise and disseminate the idea of sustainable stormwater management, promoted a philosophy of collaboration among the urban water-related industries and demonstrated the practical feasibility of pursuing this path into the future.

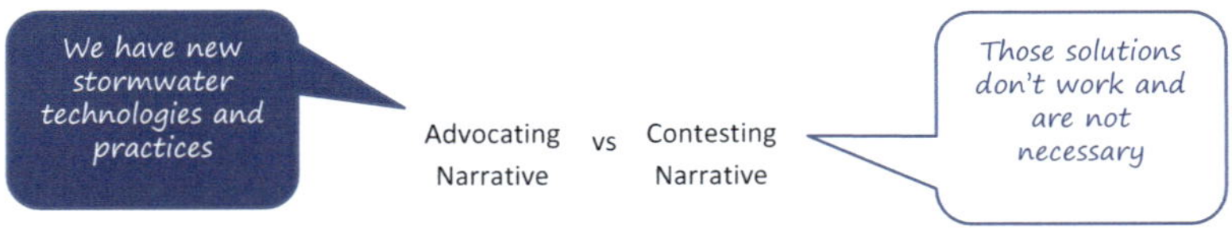

8.3.4 *Phase 4: Knowledge Dissemination (2000–2004)*

One of the key events of this phase was the first international Water Sensitive Urban Design Conference. For the first time, this conference brought together international stakeholders involved with implementing sustainable water management to exchange their experiences. Meanwhile, national and state-based best practice guidelines for urban water management were developed, and the CRC for Catchment Hydrology created a computer-based decision support tool called MUSIC to simplify the adoption of urban stormwater quality management measures in Australia (*Model for Urban Stormwater Improvement Conceptualisation*).

One of the most defining activities of this phase was political lobbying. The earlier success of the water sensitive residential development at Lynbrook Estate led water management champions to lobby the state land developer to apply water sensitive urban design principles and technologies to the Docklands, an iconic redevelopment in the centre of Melbourne. In local municipalities around Melbourne, advocates from within councils also lobbied their organisations to set up trials and encouraged other municipalities to do the same. At a more strategic level, a number of local champions took the opportunity of a pending state election to lobby the state opposition party to take a leadership role on stormwater to garner the support of the community and private sector land developers. When the opposition was voted into government in 2000, this move resulted in the establishment of an AU$22 million state fund to develop stormwater management plans and fund capacity building for stormwater professionals. This, in turn, led to the rapid increase of stormwater management actions around Melbourne, as well as to the establishment of a state Stormwater Advisory Council which championed policy innovation at a senior decision-making level rather than at a technical/operational level as previous policy groups had done.

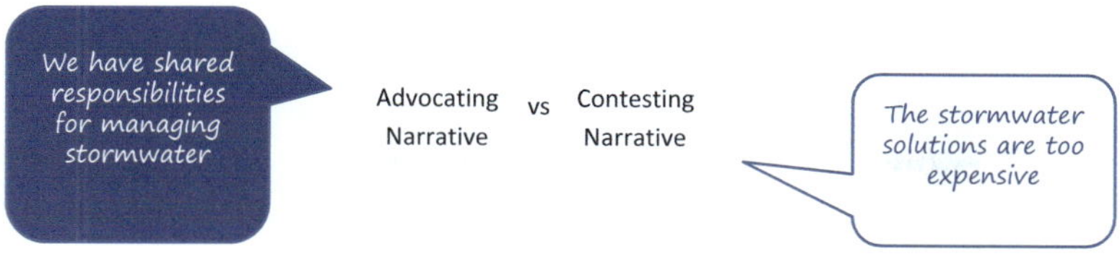

8.3.5 *Phase 5: Policy and Practice Diffusion (2004–2010)*

During this latter part of the acceleration phase, advocates focused on creating formal policy documents and pursuing regulatory change. Multi-stakeholder partnerships were now formalised through signed agreements that set out the responsibility of each organisation for improving waterway health. One of several important documents produced during this period included a planning framework that identified statutory opportunities to influence the implementation of water sensitive urban design. This was adopted by Melbourne Water and integrated into an offset strategy that provided an incentive for land developers to incorporate stormwater treatment into their planning rather than paying a charge for nutrient run-off once a development was completed.

The focus of the informal network of champions at this time was to pursue amendments to the Victorian Planning Provisions so that stormwater quality targets became a regulatory requirement. The network also encouraged broader reforms such as the introduction of an innovative offsets scheme. Over this time local councils were also experimenting more confidently with different approaches to improving stormwater management. Despite the substantial progress made with water management, a setback occurred when a persistent and severe drought resulted in a political shift back to the narrower, more traditional water supply focus. As the state government began to redirect resources to controversial options such as seawater desalination and limited wastewater recycling, the champions adapted their lobbying approach by promoting harvested stormwater as a better source of supply, thereby also implicitly promoting stormwater quality treatment.

An opportunity for strong media engagement on stormwater management arose when a kayaker fell ill after falling into the Yarra River in 2006. This event happened to coincide with the event of the 4th Water Sensitive Urban Design Conference being held in Melbourne and the ensuing media attention led to a further AU$22 million funding allocation to capacity building of municipal council staff for stormwater treatment as well as new on-ground water management projects in four high-profile municipalities.

8.3.6 *Phase 6: Embedding New Practice (2011–Present)*

By 2011, many of the early sustainable water management champions had now reached senior roles in their departments and organisations and worked to influence the political opposition party for better management of the total water cycle. This advocacy role became an advisory one when the opposition came to power and convened an independent Ministerial Advisory Council to the new government, comprising some key champions. The government also formed a new agency, the Office of Living Victoria, to implement sustainable water management objectives and facilitate interactions between the bridging organisations focused on science, policy and capacity building. While this agency had a limited lifespan, its responsibilities and mandate continue as part of the state Government Department of Environment, Land, Water and Planning.

A further important development of this phase was the launch of the CRC for Water Sensitive Cities with an explicit focus partly on stormwater harvesting and treatment. This CRC replaced the two earlier CRCs, which had reached the end of their funding terms in 2005. The new CRC's objective is to help Australian cities and towns to become more water sensitive by improving their urban water systems using tools and technology being developed through the CRC's research.

In order to become fully stabilised as the new standard way of operating, all 38 municipalities across Melbourne need to fully commit to improved storm-water quality management by directing appropriate levels of resourcing to water sustainability and better integrating stormwater quality management with other council processes. However, the strong conceptual, technical and operational links between stormwater quality and stormwater harvesting are ensuring that this emergent niche will contribute actively to making sustainable urban stormwater management the norm.

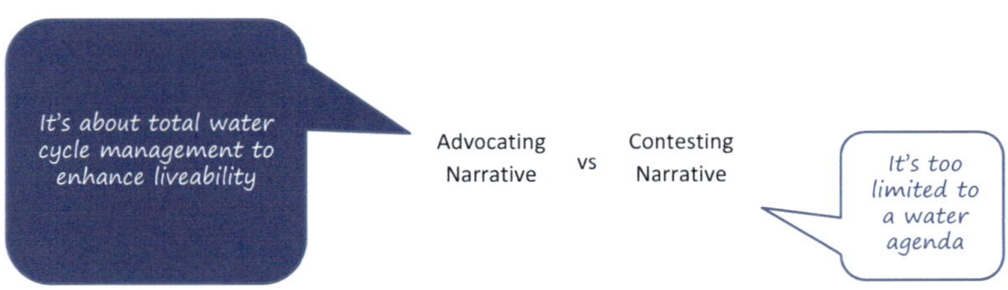

8.4 Navigating the Transition to a Water Sensitive City

In the early phases of Melbourne's transition to a Waterways City, the dominant advocating narrative reflected a realisation that stormwater pollution is causing poor waterway health, while the dominant contesting narrative rejected this assertion. By the end of the transition, the advocating narrative was that improved stormwater management helps deliver enhanced liveability outcomes, while the contesting narrative continued its challenge by questioning the value of improved stormwater management by diminishing its ability to address society's goals. These shifts highlight the value of narratives as a useful indicator of the dominant perceptions and therefore transition phase, as their evolving structure reflects the power of advocates of a new practice growing while the contesting voices that challenge the necessity or suitability of new practices shrink in relative power. In drawing attention to the dominant conversation unfolding in a sector, narratives can usefully reveal transition progress and also encourage reflexivity among practitioners to better understand their system by considering which voices speak the loudest.

More detailed analysis of the key strategies and processes that occurred within each of these phases during Melbourne's transition to a Waterways City identified five domains of change that can be used as indicators to support a more comprehensive diagnosis of the transition status and opportunities. These domains – actors, bridges, knowledge, projects and tools – influenced and organised the rules for implementing improved stormwater management practice in Melbourne. Their characteristics evolved over the course of the transition, with new dimensions added as the new practices moved from being novel to mainstream.

- *Actors:* Individual people, organisations and networks that are involved in or engaged with water management. Over time, actors involved expand from being a focused group of activists, science leaders who work to expose and define key issues, then a network of technical specialists and then to a broader coalition of policy-makers and decision makers who have the power to institutionalise the new practice. This reflects the emphasis in transitions literature on the importance of strengthening agency through shadow networks and leadership to drive transitional change (e.g. Huitema and Meijerink 2010; Olsson et al. 2006).
- *Bridges:* Organisations, structures and processes that facilitate collaborations across science, policy and industry spheres. The role of bridging organisations and mechanisms in enabling transitions is well explored in the literature (e.g. Berkes 2009; Folke et al. 2005). In the early phases of a transition, bridges help to deepen understandings of the problem and at later phases assist with translating the new practice into action. As such, the number of bridging organisations and mechanisms grows over the course of the transition.
- *Knowledge:* Scientific understanding of the problem and the potential solutions, along with contextualised knowledge informed by local research activi-

ties. Knowledge is initially developed through fundamental science and pilot-scale investigations, with the later phases involving more applied research and capacity-building initiatives. This reinforces the message of transition scholars that different types of knowledge are critical for driving changes towards sustainability (e.g. Beierle 2002; Hedelin 2007).

- *Projects:* Experiments, demonstration and focus projects to test the viability of new technologies or approaches. Experimentation and demonstration projects are well understood in the literature to be important catalysts for transitional change (e.g. Huitema et al. 2009; Vreugdenhil et al. 2010). Projects may include the development of scientific prototypes, demonstration projects that serve as proof of concept of a new approach and large-scale field applications that build trust and sector-wide capacity.
- *Tools:* Administrative and practice tools such as legislative and regulatory instruments, market mechanisms, models and best practice guidelines to help embed and structure the new practice. More recent transitions literature explores how innovations are embedded through tools to support knowledge diffusion and formal regulatory and policy structures (e.g. Kivimaa and Kern 2016; Rogge and Reichartdt 2016; Howlett and Rayner 2007). Early tools enable and support innovative approaches through practice guidance, while the tools in later phases focus more on compliance and enforcement.

Drawing on Melbourne's transition from a Drained City to a Waterways City, the results offer valuable insight into the possible dynamics of city transitions towards the Water Sensitive City more generally and in particular the changes across the five domains of change that are conducive to enabling the required practice change. A key lesson is that mutually reinforcing shifts are needed across all five domains of change for a transition to occur within a particular problem context. For Melbourne to build on the momentum of its stormwater management transition, its broader transition to a Water Sensitive City would likely maintain its actors and bridges (which are now sufficiently diverse to be relevant beyond storm water), while developing new knowledge, projects and tools focused on the whole water cycle and integration with the urban landscape.

Table 8.1 summarises the dominant narratives and indicators across the domains of change for each of the six transition phases. Reading the table from left to right highlights the developments that occurred across each domain of change during the different transition phases, while reading it from top to bottom describes the evolving nature of each domain of change over the course of a transition.

While Table 8.1 needs application and testing in other city contexts, it proposes a framework that could be used to understand and explore a city's current phase in the transition to a Water Sensitive City and potentially for urban sectors beyond water. In identifying generic indicators of change across the five domains, the framework can be used by other Australian cities to assess currently unfolding water sector transitions. Water challenges remain a key issue across Australia, with many local

Table 8.1 Indicators of change in urban water transitions

	Dominant narratives		Domains of change				
Transition phase	Advocating message	Contesting message	Actor's: Key networks of individuals and organisations	Bridges: Organisations, structures and processes for coordination and alignment	Knowledge: Research, science and contextualised knowledge	Projects: Experiments, demonstrations and focus projects	Tools: Legislative, policy, regulative and practice tools
1. Issue emergence	Problem	No problem	Issue activists	N/A	Issue discovery	High-profile scientific studies	N/A
Mid-1960s–1989			Citizens, media		Fundamental science on waterway health	Port Phillip Bay Environmental Study	
2. Issue definition	Cause	Problem and cause contested	Science leaders	Science-industry	Cause-effect	Laboratory-based and scientific solution prototypes	N/A
1990–1995			Scientists, engineers	CRCs, Melbourne Water	Stormwater pollution impacts, new technologies for removing pollution	Laboratory testing of new technologies	
3. Shared understanding and issue agreement	Solution	Solution contested	Technical solution coalition	Science-industry-policy	Basic technological solutions	Minor scientific field demonstrations	Draft best practice guidelines

(continued)

Table 8.1 (continued)

	Dominant narratives		Domains of change				
Transition phase	Advocating message	Contesting message	Actor's: Key networks of individuals and organisations	Bridges: Organisations, structures and processes for coordination and alignment	Knowledge: Research, science and contextualised knowledge	Projects: Experiments, demonstrations and focus projects	Tools: Legislative, policy, regulative and practice tools
1996–1999			Scientists, engineers, planners, land developers and local government	CRCs, Melbourne Water, Lynbrook Estate, stormwater policy committee	Treatment wetlands	Lynbrook Estate demonstration	Best practice guidelines for stormwater quality management
4. Knowledge dissemination	Responsibility	Solution not viable	Informal policy coalition	Science-industry-policy-capacity building	Advanced technological solutions	Major scientific field demonstrations	Best practice guidelines, targets
2000–2004			Scientists, engineers, planners, land developers, local government, policymakers	Stormwater Advisory Council, international WSUD conference, political lobbying	Biofiltration systems	Docklands demonstration	Pollution reduction targets

5. Policy and practice diffusion	Solution works	Solution not viable	Policy and decision coalition	Science-industry-policy-capacity building	Modelling solutions, capacity building	Numerous industry-led field experiments	Legislative amendments, market offsets, national best practice guidelines, regulatory models
2004–2010			Scientists, engineers, planners, land developers, local government, policymakers, senior decision-makers	Formalised multi-stakeholder partnerships	MUSIC model development, capacity building programme 'Clearwater'	Local government experimentation	Planning framework, developer offsets for nutrient run off
6. Embedding new practice	Solution delivers prosperity and liveability	Solution is insufficient for meeting a wider set of needs	Multi-agency coalition	Formalised institution	Next research agenda	Standard practice	Political mandate, coordinating authority, comprehensive regulatory models and tools
2011–present			Executive level champions across key organisations	Office of Living Victoria, DELWP	Managing storm water as part of the whole water cycle	Integration into council processes	Ministerial attention, DELWP, MUSIC, practice guidance

and state governments looking to move towards the Water Sensitive City vision. This framework provides a means of assessing progress towards this goal and can also be used to inform investment and the development of a strategic transition plan going forward. The framework can also be used to examine water transitions in countries beyond Australia and may also provide a basis for the development of a similar framework tailored to other sectors.

By identifying the dominant advocating and contesting narratives that are heard by different actors across the sector, and diagnosing which of the domains of change are already present, key gaps may be revealed. For example, if the framework indicates that the city in question is currently in Phase 4 (Knowledge Dissemination) and all indicators except the *Informal policy coalition* are present, further progress would be best accelerated by focusing on the enabling conditions that help to build that missing coalition to further advance water sensitive management practices towards Policy and Practice Diffusion. Considering the enabling conditions most relevant to particular phases of change provides step-by-step guidance on how to build up a foundation to support a broader transition. This type of guidance can provide strategists, policymakers and practitioners with clear direction on how to most efficiently and productively invest resources as they navigate their city's water sensitive transition.

8.5 Conclusion

In exploring Melbourne's stormwater quality management transition, this chapter reveals the complexity of urban transition processes and the challenge on realising change on the ground. To simplify understanding of this process, and provide guidance for policymakers and change advocates seeking to realise change in their own cities, we have presented a framework that can help identify the types of strategic actions that are needed and their implementation priorities.

The framework in Table 8.1 can be used to qualitatively benchmark the current state of a city's transition towards water sensitive practices, facilitating an integrated and detailed understanding of a city's water management practices within its real-world context. Key issues to consider are what strategic positioning requirements are needed to strengthen the advocating narratives and challenge the contesting narratives, as well as what types of strategic actions will help develop the actors, bridges, knowledge, projects and tools that are required for advancing the transition progress. Ultimately, transition processes are complex, and the case of Melbourne makes clear that ongoing commitment, monitoring and investment are needed in order to guide change in desirable directions and successfully realise change on the ground.

References

Beierle T (2002) The quality of stakeholder-based decisions. Risk Anal 22(4):739–749

Berkes F (2009) Evolution of co-management: role of knowledge generation, bridging organizations and social learning. J Environ Manag 90:1692–1702

Brown R, Farrelly M (2009) Delivering sustainable urban water management: a review of the hurdles we face. Water Sci Technol 59(5):839–846

Brown RR, Keath N, Wong THF (2009) Urban water management in cities: historical, current and future regimes. Water Sci Technol J Int Assoc Water Pollut Res 59(5):847–855. doi:10.2166/wst.2009.029

Brown R, Farrelly M, Loorbach D (2013) Actors working the institutions in sustainability transitions: the case of Melbourne's stormwater management. Glob Environ Chang 23(4):701–718. doi:10.1016/j.gloenvcha.2013.02.013

De Haan J, Rotmans J (2011) Patterns in transitions: understanding complex chains of change. Technol Forecast Soc Chang 78(1):90–102. doi:10.1016/j.techfore.2010.10.008

De Haan FJ, Rogers BC, Frantzeskaki N, Brown RR (2015) Transitions through a lens of urban water. Special issue editorial. Environ Innov Soc Trans 15:1–10

Ferguson BC, Frantzeskaki N, Brown RR (2013a) A strategic program for transitioning to a water sensitive city. Landsc Urban Plan 117:32–45

Ferguson BC, Brown RR, Frantzeskaki N, de Haan FJ, Deletic A (2013b) The enabling institutional context for integrated water management: lessons from Melbourne. Water Res 47:7300–7316

Folke C, Hahn T, Olsson P, Norberg J (2005) Adaptive governance of social–ecological systems. Annu Rev Environ Resour 30(1):441–473

Hedelin B (2007) Criteria for the assessment of sustainable water management. Environ Manag 39(2):151–163

Howlett M, Rayner J (2007) Design principles for policy mixes: cohesion and coherence in 'new governance arrangements. Policy Soc 26:1–28

Huitema D, Meijerink S (2010) Realizing water transitions: the role of policy entrepreneurs in water policy change. Ecol Soc 15(2):26

Huitema D, Mostert E, Egas W, Moellenkamp S, Pahl-Wostl C, Yalcin R (2009) Adaptive water governance: assessing the institutional prescriptions of adaptive (co-) management from a governance perspective and defining a research agenda. Ecol Soc 14(1):26

Kivimaa P, Kern F (2016) Creative destruction or mere niche support? Innovation policy mixes for sustainability transitions. Res Policy 45:205–217

Markard J, Raven R, Truffer B (2012) Sustainability transitions: an emerging field of research and its prospects. Res Policy 41(6):955–967. doi:10.1016/j.respol.2012.02.013

Olsson P, Gunderson LH, Carpenter SR, Ryan P, Lebel L, Folke C, Holling CS (2006) Shooting the rapids: navigating transitions to adaptive governance of social-ecological systems. Ecol Soc 11:18

Pahl-Wostl C, Sendzimir J, Jeffrey P (2009) Resources management in transition. Ecol Soc 14(1):46

Pierson P (2004) Politics in time: history, institutions and social analysis. Princeton University Press, Princeton

Rogers BC, Brown RR, de Haan FJ, Deletic A (2015) Analysis of institutional work on innovation trajectories in water infrastructuresystems of Melbourne, Australia. Environ Innov Soc Trans 15:42–64

Rogge K, Reichartdt K (2016) Policy mixes for sustainability transitions: an extended concept and framework for analysis. Res Policy 45(8):1620–1635

Rotmans J, Kemp R, Van Asselt M (2001) More evolution than revolution: transition management in public policy. Foresight 03(01):15–31

Smith A, Stirling A, Berkhout F (2005) The governance of sustainable socio-technical transitions. Res Policy 34(10):1491–1510

Vreugdenhil H, Slinger J, Thissen W, Ker Rault P (2010) Pilot projects in water management. Ecol Soc 15:13

Walker W (2000) Entrapment in large technology systems: institutional commitment and power relations. Res Policy 29(7–8):833–846. doi:10.1016/S0048-7333(00)00108-6

Wong THF, Brown RR (2009) The water sensitive city: principles for practice. Water Sci Technol 60(3):673–682

Chapter 9
Transitioning the Greyfields

Peter W. Newton

Abstract In October 2015 *Plan Melbourne Refresh* (DELWP, Plan Melbourne refresh discussion Paper. Department of Environmental Land Water and Planning, Melbourne, 2015) recognised 'greyfield precinct renewal' as a significant new model for more intensified and sustainable 'urban' redevelopment in the established, ageing inner- and middle-ring 'suburban' areas of the Melbourne Metropolitan Region. This chapter documents the critical phases of this 6-year 'shadow urban transitions process' (*Greening the Greyfields* project). Framed within a multilevel Transition Management schema, it addresses a set of challenging ('landscape') factors necessitating urban transformation together with the 'regime' barriers that are blocking more effective and sustainable forms of urban retrofitting of established low-density suburbia – at precinct scale. The transition process required articulation of a new model for greyfield precinct regeneration, necessitating 'niche' innovation in several critical arenas within the property development process that involve multiple stakeholders – government, industry and local communities. These innovations encompass new digital, governance, community engagement, codesign and planning instruments developed to support implementation of the new model for greyfield precinct regeneration.

Keywords Urban transition • Precinct regeneration • Greyfields • Medium-density housing

9.1 Introduction

Australia's fast-growing cities are outstripping the capacity of city planners to shape their development in a sustainable manner. Projections to 2061 for the four largest capital cities indicate that they will continue to be the principal focus for the nation's growth, absorbing at least two thirds of total population, with annual rates of growth unmatched in comparable OECD countries (IA 2015). By then, Melbourne

P.W. Newton (✉)
Institute for Social Research, Swinburne University of Technology, Melbourne, VIC, Australia
e-mail: pnewton@swin.edu.au

T. Moore et al. (eds.), *Urban Sustainability Transitions*, Theory and Practice of Urban Sustainability Transitions, DOI 10.1007/978-981-10-4792-3_9

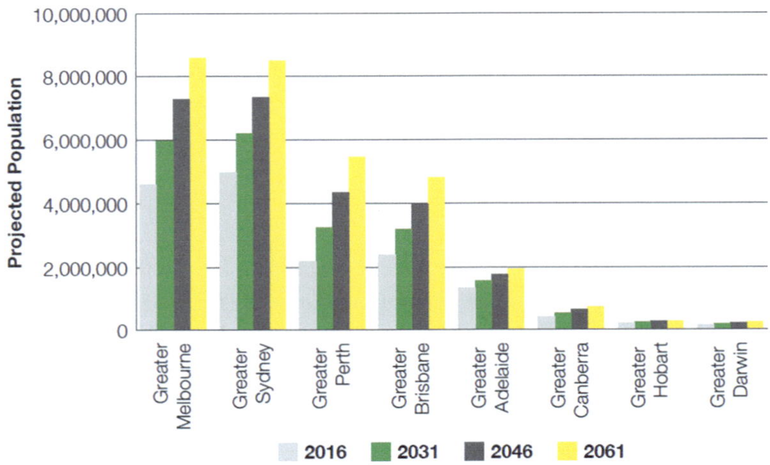

Fig. 9.1 Projected population growth in Australian capital cities – 2016, 2031, 2046 and 2061 (Source: Infrastructure Australia, Australian Infrastructure Audit – Supplementary Reports – Population Estimates and Projections, Commonwealth of Australia, Sydney, 2015)

is forecast to be the largest city (8.6 million, double present population), followed by Sydney (8.5 m), Perth (5.5 m) and Brisbane (4.8 m; DIRD 2015; Fig. 9.1). Whether continued high levels of urban growth exert positive or negative impacts on the future economic, social and environmental performance of Australian cities will depend on how well they are planned. However, new metropolitan planning policies in Australia must depart from those that have delivered some of the world's least sustainable cities as measured by their ecological and carbon footprints (Newton 2012). They must redirect population growth inwards and upwards, rather than outwards.

Successive strategic metropolitan plans for the major cities in Australia over the past 20 years have attempted to slow suburban sprawl by introducing a number of 'compact city' policies, including growth boundaries, protected green wedges, zoned activity centres and major transport corridors where higher-density redevelopment was to be encouraged, and more recently urban infill targets for new housing ranging between 40 and 70% of new construction. That growth boundaries have been redrawn (further out), green wedges compromised, activity centres and transport corridors failing to act as the magnets for more intensified urban redevelopment as 'planned' and (net) housing infill struggling to reach 50% of new builds (Newton and Glackin 2014) is due to a continued failure on the part of governments to appropriately plan for growth. Gleeson et al. (2012) provide insightful accounts of the planning and implementation 'deficits' that have characterised this recent era of metropolitan development in Australia.

Faced with the challenge of increasing housing supply and variety to meet the needs of a rapidly growing and changing population, a new logic for urban development is required *if* critical transitions are to occur in Australia's major cities. These include:

- Transition to *more compact*, transit-oriented and walkable cities
- Transition in property development with *urban infill* assuming a dominant role in the supply of *medium-density housing*, in addition to the recent boom in high-density inner city and CBD apartment construction
- Transition from an essentially low-density suburban city form to an *urban form*, more characteristic of European cities that deliver urban liveability more sustainably than those in Australia and North America
- Transition to a more *sustainable built environment* whose direct and indirect consumption of energy, water, housing space, urban travel and generation of wastes is within the planetary boundaries of a global ecosystem required to support an estimated 9 billion population by 2050, 75% of whom are expected to be urban (Newton and Doherty 2014)

This new logic needs to embrace the regeneration of the extensive *greyfield arenas* of Australian cities. The greyfields have been defined by Newton (2010) as those areas where the building stock is ageing physically and technologically and is environmentally poor performing. The value of property in the greyfields lies predominantly in the land, not the built asset. As such, the real estate is undercapitalised and is prime for redevelopment. The greyfields are located in the established middle suburbs, between the CBD (and inner city housing markets) and the more recently developed greenfield suburbs. They are well located with respect to public transport access, jobs and specialist health services and tertiary education. Their retrofit potential has been identified as high (Newton et al. 2011; MAC 2015). However, they differ fundamentally from their counterpart arenas of *greenfields* and *brownfields* in that they are occupied, thereby introducing a much greater level of contestation and complexity into decision-making related to the nature and scale of their redevelopment.

These three urban development arenas (greenfields, brownfields and greyfields) occupy distinctive locations within the urban geography and fabric of cities, ranging from the inner walking and transit city to the outer suburban car-dependent city (see Fig. 9.2 and Table 9.1). Models and processes associated with the urban development of greenfields have been well established for decades in all Australian cities, with master planned precincts being among the more recent products to emerge. However, they continue to perpetuate many of the problematic social, economic and environmental attributes of existing outer suburban areas (low density, poor access to jobs, services and public transport). Models for brownfield development have emerged more recently as a result of the federally funded Building Better Cities (BBC) Program that commenced in 1991 (Neilson 2008). State governments have continued to identify brownfield areas as strategic urban renewal precincts. Their capacity to supply more than a small percentage of future new housing construction is limited, however (Newton and Glackin 2017).

As the third arena for property development in cities, government planning strategies for the greyfields have to date centred on an attempt to encourage

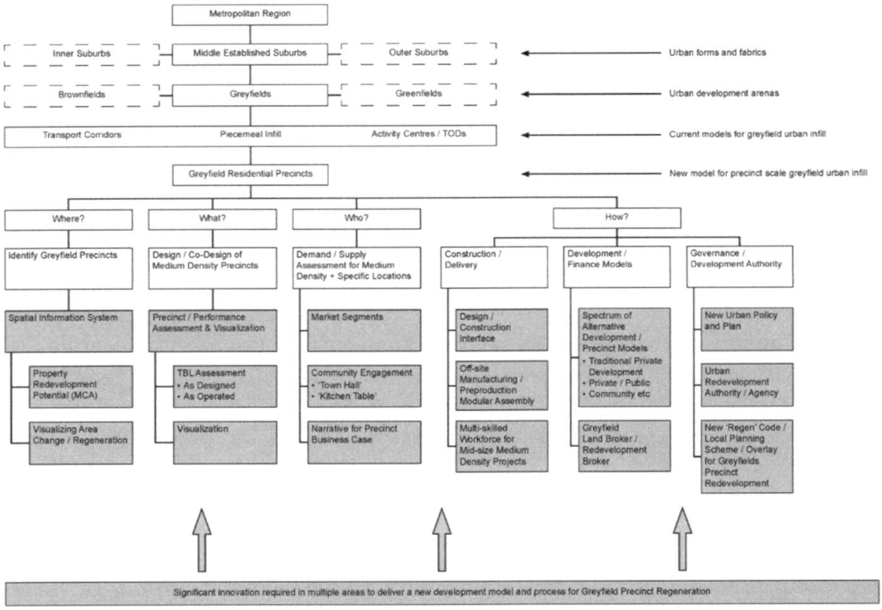

Fig. 9.2 Innovation arenas for establishing greyfield renewal precincts (Source: Modified from Newton et al. 2011)

significant urban infill and intensified redevelopment in three spatial contexts: activity centres, transport corridors and piecemeal knock-down rebuild of individual properties (again see Fig. 9.2). An extensive description, assessment and critique of these three current greyfield infill redevelopment strategies is found in Newton (2010), Newton and Glackin (2014, 2015) and Newton et al. (2011).

To summarise, most greyfield infill housing redevelopment is currently occurring as suboptimal, fragmented, single lot knock-down rebuild development that is largely located outside of local and state government designated redevelopment areas. While activity centre and transport corridor-centred redevelopment are both considered necessary components of strategic metropolitan planning policies designed to facilitate the emergence of more compact cities and halt sprawl, they are not of themselves sufficient. Furthermore, they are not contributing the sustainability 'dividend' required for the *regeneration* of Australian cities, defined here (after Girardet 2015 and Birkeland 2014) as development capable of dramatically reducing the level of resource consumption that is required to create and operate the built environment and where possible renew and restore the ecosystem support of cities. It will be argued here that eco-positive development is most effectively delivered via addition of a *precinct-scale* urban renewal planning policy for the greyfields.

The remainder of this chapter documents the critical phases of a 6-year 'shadow urban transitions process' (*Greening the Greyfields*) project that advances a new urban planning concept for attempting more regenerative, precinct-scale, urban

Table 9.1 Urban land-use arenas for future urban planning and development

More compact cities	
Brownfields	
Brownfield precinct regeneration	Small- and large-scale brownfield industrial and abandoned retail precincts; high-density redevelopment; mixed use; public transit oriented
Greyfields	
Greyfield piecemeal/fragmented residential redevelopment	Replacement of single dwellings (typically detached) with 2–4 townhouses
Greyfield residential precinct regeneration	Low-rise medium-density; mixed use; regenerated water and energy systems; community-centred; mixed typologies, floor areas and layouts, etc.
Activity centre development	Higher-density redevelopment (residential, commercial) centred on state government designated activity centres; in some instances with related TOD enhancements
Transport corridor development	Higher-density residential and commercial development along major transport corridors, e.g. principal arterials, tram routes, light rail
More urban sprawl	
Greenfields	
Business-as-usual greenfields	Low-density predominantly detached housing; car dependent; little or no mixed use
Smart greenfields	Mixed uses and moderate densities (mixed housing types); energy and water efficient; walkable with public transport access

transformation in the greyfields (Newton 2010). Framed within a multilevel Transition Management schema, the form and scale of greyfield redevelopment envisaged here address a set of challenging ('landscape') factors necessitating urban transformation together with the 'regime' barriers that are blocking more effective and sustainable forms of urban retrofitting of established low-density suburbia – at precinct scale. The transition process required articulation of a new model for greyfield precinct regeneration, necessitating 'niche' innovation in several critical arenas within the property development process that involve multiple stakeholders – government, industry and local communities. These 'innovation arenas' encompass the new digital, governance, community engagement, codesign and planning, building finance and delivery instruments required to support implementation of the new model for greyfield precinct regeneration (again see Fig. 9.2). In October 2015 *Plan Melbourne Refresh* (Government of Victoria DELWP 2015) recognised 'greyfield precinct renewal' (GPR) as a significant new model for more intensified and sustainable urban redevelopment in the established, ageing inner- and middle-ring suburbs of the Melbourne Metropolitan Region. The following sections outline the applied transitions research and development that underpins this new greyfield redevelopment model. The *multilevel process* provides a useful *theoretical-descriptive framework* for representing the context and dynamics within which urban innovations emerge, including the barriers that need to be overcome in order to deliver significant systemic change. *Transition Management* represents the *methods* and sets of participatory processes by which a model for change is identified, examined and implemented – in this instance GPR.

9.2 The Challenge of Greyfield Precinct Renewal: Application of Transition Theory's Multilevel Process (MLP) Framework

A regenerative retrofitting of Australian cities requires innovation and change that is radical, transformative and potentially disruptive. Sociotechnical transition theory has emerged in recent years in an attempt to provide greater conceptual coherence to the complex set of processes involved in the challenging transition to sustainable development (Geels 2011; Grin et al. 2010). The processes are *multilevel* and interconnected and revolve around three major constructs: the landscape, the regime, and the niche innovation – applied here in a contemporary (as distinct from the more common historical) urban context.

9.2.1 Landscapes

'Landscapes' represent and reflect the prevailing sets of exogenous and endogenous conditions and processes in current urban environments, such as those that feature in regular *State of Environment* reports and *State of Australian Cities* reports as well as contemporary academic reviews (e.g. Newton and Doherty 2014). They are the products of existing and prior local regimes and practices as well as global forces and represent the context within which niche innovations emerge and attempt to penetrate the market. Each urban innovation – as identified in Fig. 9.2 (grey boxes) – has a role to play in helping progress a transition towards more sustainable urban development (Newton and Bai 2008). The principal urban 'landscape' problems in Australia's major cities relevant to precinct-scale urban infill innovations are:

- A shortage of housing supply capable of matching demand (NHSC 2013), a rapid growth in residential property prices to a position where Australian capital cities lead 26 other global housing markets as tracked by *The Economist* (2015) and a continuous decline in housing affordability (Wood and Ong 2015). Much of the higher-priced housing is now concentrated in the inner and middle suburbs of Australia's largest cities – reflecting processes of economic restructuring, re-urbanisation and gentrification focusing on the inner city.
- Continued dependence on greenfield development to accommodate the majority of new housing constructed in Australian cities, with most cities currently

struggling to reach 50% of new housing as infill, ensuring that there will be continued urban sprawl together with its established negative consequences (Trubka et al. 2010).

- An extended period of house price inflation above that of wages and cost of living forcing lower-income households to cheaper, less accessible and less productive locations in the car-dependent outer suburbs. Suburbanisation of disadvantage has increased markedly over the past 20 years and is now clearly evident in the major cities (Randolph and Tice 2015). This socio-spatial disadvantage is multifaced and combines concentrations of low income; poor access to jobs, public transport, tertiary education and advanced health-care services; and a spatial concentration of social problems. It is weakening the social fabric as well as the productivity of Australian cities.
- A post-World War II 'suburban' urban form and density created as a result of the development of car-dependent middle and outer suburbs comprising mostly detached dwellings with floor areas that are now among the largest in the world (Shrinkthatfootprint 2016): a housing and transport 'cocktail' that has combined to produce ecological footprints for all the major Australian capital cities that are three times the world average (Newton 2012). The carbon footprints are commensurately large: in CO_2 terms, variability in the housing and transport profiles of different suburbs means that neighbourhood-scale carbon emissions can vary by as much as 50% across Australian cities (Crawford and Fuller 2011).
- Ageing urban infrastructure, linked to a national shortfall in infrastructure expenditure estimated at approximately A$100 billion – a fourfold increase in less than a decade (Watt 2014). Australian housing is also ageing physically, technologically and environmentally and when combined with the state of underpinning urban infrastructure is indicative of a need to explore optimal pathways for broader-based, precinct-scale regeneration in the established suburbs. There remains a real prospect of system breakdown in key urban infrastructures over the next 20 years for many cities, especially those facing rapid growth, coupled with impacts of climate change and within the context of a continuation of late twentieth-century neoliberal planning and urban development practice. The emergence of eco-efficient distributed energy, water, waste and transport technologies provides a regenerative pathway for urban infrastructure that meshes with precinct-scale urban development.

9.2.2 *Regimes*

The 'regime' in this context encompasses all the industries involved in property development, government at all levels and local communities, their current practices, established relationships and modes of 'doing business'. While legislative power rests with government, individual businesses and industry lobbies are powerful influences on city development as are urban communities of residents who may perceive change affecting their neighbourhood as something to be resisted. They constitute a heady mix of stakeholders in established, occupied residential areas with high redevelopment potential. The barriers to be overcome in implementing transformational change can therefore be formidable. Geels and Schot (2010) argue that major transitions are difficult because of the lock-in or path-dependent tendencies within existing sociotechnical systems (STSs). These STSs comprise complex groupings of people and practices that surround current production processes linked with city building, an established supply chain of built environment products and their related manufacturing and servicing industries and a myriad of urban infrastructure institutions (transport, water, energy, waste, construction, etc.). These STSs also embody marketers of the products, social groupings of users who are familiar with the product or service, policy makers who provide the operating regulatory frameworks, industry associations that manage the government and community interface for key industry segments. Together they form strong networks with mutual dependencies. As a rule, STSs are resistant to all but incremental change. Geels and Schot (2010) suggest, however, that windows of opportunity for transitioning existing STSs arise as a result of pressure from exogenous and endogenous forces on an existing regime (e.g. the cumulative impact that the build-up of urban problems, listed under 'landscapes', can have in causing 'ruptures' in critical parts of an urban system and their resident populations). Understanding where and why an urban system is *not* functioning effectively due to the existence of key *barriers* can provide important insights indicating how pathways for new and more innovative urban products, instruments and regimes can be established to enable transformational change.

While regimes can often be inhibitors of change, Loorbach (2010) correctly points out that regimes can also represent the enabling environment for facilitating and legitimating a transition. As such, strategies need to be developed with key stakeholders in respect of the transition arena under examination and the nature of transformation envisaged. In the case study to be outlined in the following section, the principal stakeholder is the state government department responsible for metropolitan planning, and the urban arena in focus is the greyfields. The niche innovation is 'greyfield *precinct* renewal', with the challenge for it to become a formal urban redevelopment policy in the next metropolitan strategic plan. Successful implementation could be expected to trigger transformational change in urban infill.

9.2.3 Niche Innovations

Over decades, innovations in transport, communication, energy generation, waste treatment, building, domestic and industrial appliances, etc. have transformed industries, jobs and built environments. Their capacity for urban transformational change varies as there are different horizons of urban technology innovation

(Newton and Bai 2008) as well as levels of resistance to their adoption. In this chapter, several (niche) innovation arenas have been identified as a focus for applied research in an effort to provide new models, processes, tools and instruments that are critical to greyfield precinct regeneration (GPR). GPR has been characterised as a 'horizon 3' urban innovation, given the significant sets of barriers it faces for acceptance and implementation (Newton and Bai 2008). They are arenas where new knowledge on key questions is needed: *where* to focus GPR, *what* to design for these precincts, *who* is the market and *how* to effectively deliver a GPR project (again, refer to Fig. 9.2). These niche innovation arenas were identified in the first phase of a *Transition Management* process that forms the basis of discussion in the next section.

9.3 Transition Management: From Theory to Practice and Implementation

The challenge of Transition Management (TM) is to establish a process, whereby productively disruptive niche innovations can be effectively introduced into the urban planning and development process overseen by regime stakeholders that are typically risk-averse and favour supporting short-term projects that employ business-as-usual practices. The transition arena (TA) model outlined by Loorbach (2010) provides the generic ingredients for assembling the novel instruments, processes and actors necessary for transforming policy in a specific context. They tend to be established as *virtual arenas*, comprising experts assembled as part of an applied research network to co-create the innovative new instruments and practices capable of being substituted for and/or integrated with current processes. The TA also tends to be established as a *shadow process*, comprising thought leaders who are typically not current senior employees of the key stakeholder groups in the 'urban regime', but have substantial tacit as well as research knowledge related to city development processes and are recognised for their innovative thinking and preparedness to advocate for change. This 'off-line' TA working group will at times be developing new instruments and tools that could be perceived as threatening at a political or senior bureaucratic level, hence the need for a shadow process and an associated 'incubator' environment conducive to unconstrained thinking. There is, however, a need from time to time to expose the outputs from the TA to different regime stakeholders to obtain the critical feedback needed to test and harden the new instruments, narratives and processes to a point where they can be included by government in new metropolitan planning policies and schemes. Here leadership of the shadow TA is critical, since the arena where niche innovations are being created – here related to a reshaping of the established greyfield areas of cities to achieve urban regeneration – is contentious.

Business-as-usual metropolitan planning has delivered the problematic urban 'landscapes' described earlier, with each successive strategic plan revision failing

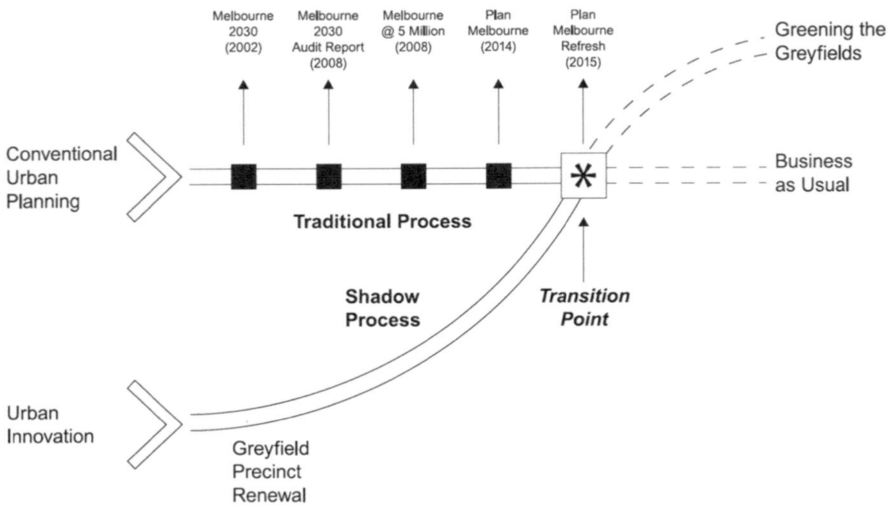

Fig. 9.3 Shadow process trajectory for *Greening the Greyfields* and Greyfield Precinct Renewal

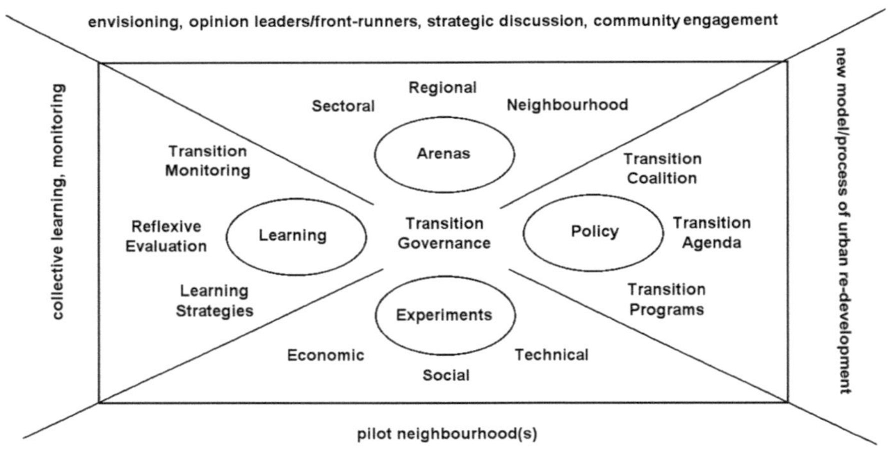

Fig. 9.4 Stages in a transition arena for Greyfield Precinct Regeneration (Source: Adapted from Loorbach 2010, p. 173)

to achieve the necessary outcomes (see Fig. 9.3). This will continue until more radical options for urban redevelopment, such as greyfield precinct renewal *and* regeneration, become part of the (Melbourne) metropolitan planning strategy. The shadow TA established in 2010 to drive this transition has been subsequently labelled *Greening the Greyfields* (*GtG* TA). TAs also comprise a number of distinctive *stages*, distinguished according to key activities undertaken in each (Fig. 9.4). *GtG* TA has been framed accordingly.

9.3.1 Stage 1: Envisioning and Scoping

In the first envisioning stage, the shadow leadership team was established under a research grant from the Australian Housing and Urban Research Institute (AHURI) and led by the author of this chapter. This enabled a series of investigative panel workshops to be undertaken over a 12-month period to expose and refine new greyfield, precinct and regenerative planning concepts to a wide spectrum of stakeholders with established reputations in key fields including urban planning (statutory and strategic), urban design, building and construction, property development, community organisation, local and state government and industry associations. Three facilitated workshops were undertaken during 2010 with 70 'frontrunner' thought leaders, generating significant feedback on such topics as: why greyfields (a new concept at that stage), needs analysis, inhibitors to regeneration of the middle suburbs, how to ensure change happens, solution pathways, investor and resident concern, creating a narrative, key delivery agencies, current planning barriers, financial challenges, etc. (see Fig. 9.5 for two of the representative Mind Maps synthesised from the workshop discussions). The results of research undertaken for this phase and the workshop outputs are reported in Newton et al. (2011). A critical output from Stage 1 of the *GtG* TA was the identification of several key areas where *innovation* was required beyond BAU before any advocated GPR policy could be expected to be endorsed in future metropolitan strategic planning. They represent the 'levers for change', and all need to be in play in order for any significant greyfield precinct regeneration to occur. These innovation arenas are identified in Fig. 9.2 and have constituted the research foci for Stage 2 of the *GtG* TA process.

9.3.2 Stage 2: New Models, Instruments, Processes and Policies for Greening the Greyfields

The innovations required for this stage of the *GtG* TA process have been categorised according to their provision of response to four questions central to implementation of greyfield precinct regeneration: *Where? What? Who?* and *How?* What follows is a précis of key outputs on these topics drawn from research and publication to date.

9.3.2.1 Where?

One of the challenges for greyfield residential precinct regeneration involves the *identification* of prospective precincts – concentrations of residential property where economic value lies predominantly in the land rather than the ageing built asset and where there are prospects of adding a community dividend from neighbourhood activation. In metropolitan Melbourne there are at least three quarters of a million dwellings where 70% or more of the value of the property is vested in the land.

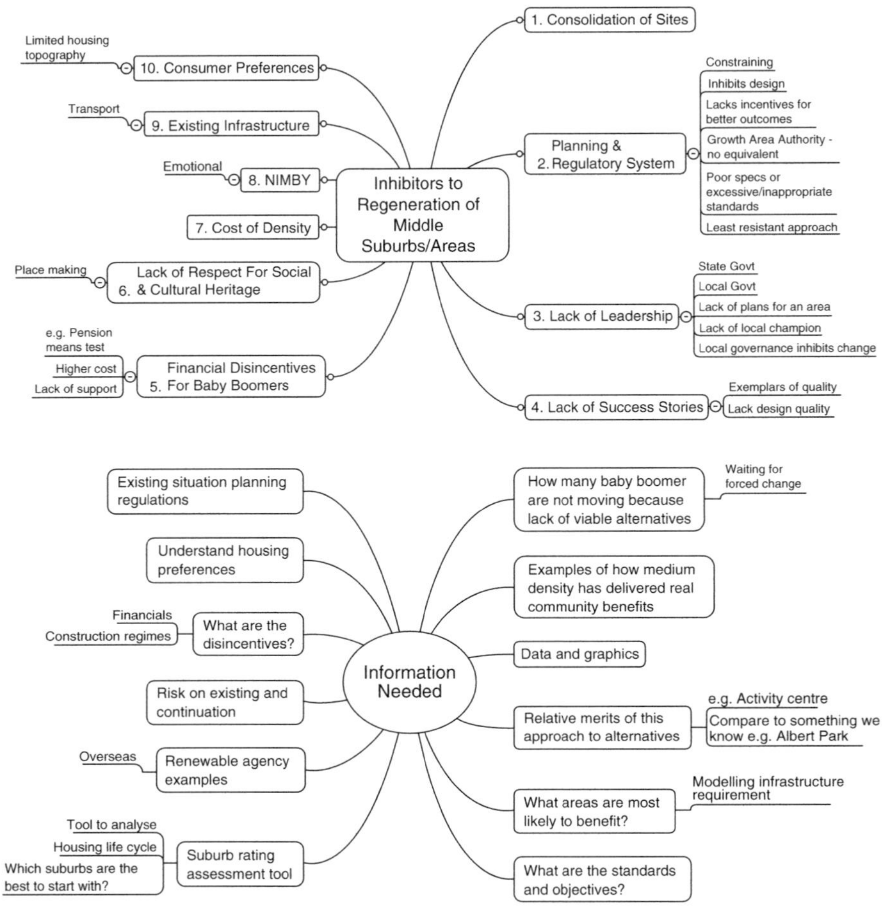

Fig. 9.5 Sample of mind maps from investigative workshops (Source: Newton et al. 2011)

A software tool (ENVISION; Newton and Glackin 2013) has been specifically developed to identify the location of properties with high redevelopment potential, employing multi-criteria analyses that are purely market based or reflect local planning priorities (e.g. proximity to schools, public transport, park, etc.) – depending on assessment criteria selected (see Fig. 9.6). ENVISION is available as an e-Research tool from AURIN and CRCSI websites (http://www.greyfieldplanning.com.au/).

9.3.2.2 What?

What to build represents the second arena for innovation in greyfield precinct regeneration. Medium density represents the design focus for *GtG* TA – there is currently a major deficit in this class of dwelling in the major cities. Precincts

Fig. 9.6 Neighbourhood Activity Centre and surrounding undercapitalised dwellings (*left*) and with potential amalgamated lots (*right*) (Source: Derived from *ENVISION*; Glackin and Newton 2016)

will be typically situated among extensive tracts of low-density detached housing that has come to define the ubiquitous 'neighbourhood character' for Australian suburbs that is enshrined in many local government residential zoning schemes and acts as a major barrier to more intensified forms of mid- to low-rise medium-density redevelopment. Poor design of much medium-density housing built to date has been identified as a major barrier to community acceptance of its diffusion more extensively throughout existing suburbs and is an issue that governments are attempting to address: 'while there is an undeniable imperative to increase densities... there is a need to ensure that new developments are well designed, are respectful of the environment and contribute positively to the existing built form' (NSW Planning 2011, p4). The urban design professions have the capacity to address this by injecting the innovation currently available into the design of new buildings and precincts. Urban design for precinct renewal needs to create, in addition to dwelling yield, a new and attractive character for the established suburbs that is more 'urban' in nature as a result of the substitution of well-designed medium-density housing for detached housing. The precinct redesign also needs to be regenerative, by the introduction of renewable energy and integrated (storm water, wastewater) water systems at building and precinct scales, food composting and a reconfiguration of the land previously devoted to the automobile to incorporate rain gardens and urban forests and pedestrian-friendly spaces – enhancing urban microclimate and walkability of the neighbourhood (see Fig. 9.7). Precinct-scale property development can now be better integrated with the emergence of distributed technologies related to energy (Newton and Newman 2013), water (CRC for Water Sensitive Cities 2016), waste (Waller et al. 2017) and mobility (Newton et al. 2017b).

For GPR to be accepted as a new planning instrument and property development option, however, critical neighbourhood engagement tools and processes associated

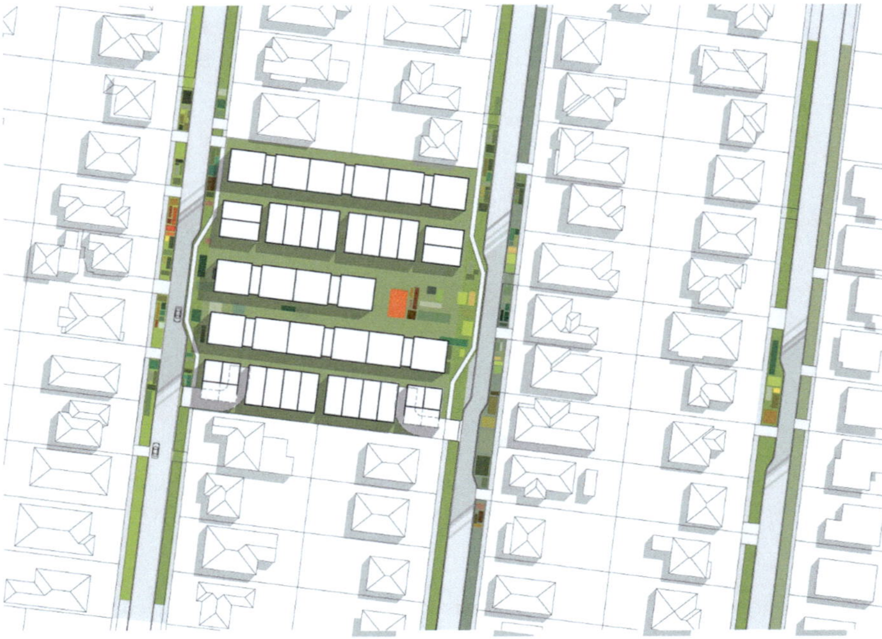

Fig. 9.7 Regenerative design of greyfield precincts (Source: Newton et al. 2011; Monash Architecture)

with precinct-scale medium-density housing need to be developed and trialled. Engagement processes (see How?) require a toolkit including software capable of allowing precinct designs to be *visualised* and manipulated (*codesign*) in local/neighbourhood forums in real time and *assessed* as to the performance of these designs against benchmarks that can establish the extent to which they improve upon the built environment that is being replaced. Such assessments provide an evidence base for local councils and communities to evaluate the level of dividend (benefit) that can flow from a proposed precinct redevelopment and will be instrumental in the level of support and incentive that local (and state) governments can contribute. The *GtG* TA has created an eco-efficiency performance assessment and visualisation tool for use at a precinct sketch planning stage (Envision Scenario Planner – ESP; Trubka et al. 2016; see Fig. 9.8). It represents a new generation of object-based, PIM-enabled tools that will be critical in driving an urban sustainability transition more generally (Newton et al. 2017a).

9.3.2.3 Who?

A large and identifiable market has been identified for new medium-density housing in the established suburbs of Australian cities. Recent surveys (Kelly

Fig. 9.8 Envision Scenario Planner (ESP) for performance assessment of a precinct sketch design (Source: Trubka et al. 2016)

2011; Newton and Glackin 2015) indicate that there is significant demand for 'urban' as well as 'suburban' living. The supply side of the housing market is beginning to respond as revealed by recent trends in dwelling approvals, where units (mostly high-rise apartments above four storeys) are now accounting for more than half of all approvals in Sydney and Melbourne (Kusher 2015). There remains a significant deficit in and demand for new medium-density greyfield precinct projects comprising 2–4 storey dwellings (Newton and Glackin 2014). Meanwhile, there is massive under-occupancy of detached housing among the 55–80-year old cohort in Australian cities – groups of household currently provided with little option than to age in place or move to a retirement village or aged care hostel. *Citizen-led* GPR represents a prospective new model, whereby neighbouring property owners in the greyfields can capture significant windfall profits by amalgamating their holdings to provide for precinct-scale redevelopment. How to achieve this represents the principal challenge for *GtG* TA.

9.3.2.4 How?

Innovation is required on several fronts (again see Fig. 9.2):

Community Engagement and Land Assembly This is fundamental at multiple levels. At metropolitan level, this involves ensuring there is a clear metanarrative developed by state government that can be broadcast as well as advocated by the major media outlets, clearly outlining the vision for the future city as well as conveying the principal pillars of any new metropolitan strategy (especially the more contentious elements such as urban infill and intensification). The extent to which such an awareness and concern can be engendered with the citizens regarding the critical problems confronting the city that necessitates some radical planning response represents an important step towards achieving change. The privileged 'I'm alright, Jack' syndrome among property owners in the established middle suburbs remains an obstacle to be overcome.

At municipal level, structuring the 'Town Hall' meetings provides the principal platform for local governments to outline their visions and strategies for future local development. A partnership with the *GtG* TA enabled the City of Maroondah (2016) to create its first *Municipal Housing Strategy 2016* as a vehicle for engagement, building community support for change and reducing NIMBYism. Housing strategies are now required from all municipalities within the Melbourne metro region to help better align state and local government planning policies.

At neighbourhood level, a new class of change agent is required capable of being a trusted and knowledgeable *broker* for 'Kitchen Table' meetings among neighbours interested in exploring property amalgamation in a newly emerging citizen-led property redevelopment process (see Fig. 9.9). This is a new phenomenon in the established suburbs of Australia's largest cities and requires a range of professionally validated instruments to be assembled as a toolkit for brokers: legal (e.g. process for consolidating titles), financial (e.g. economic feasibility, tax liability),

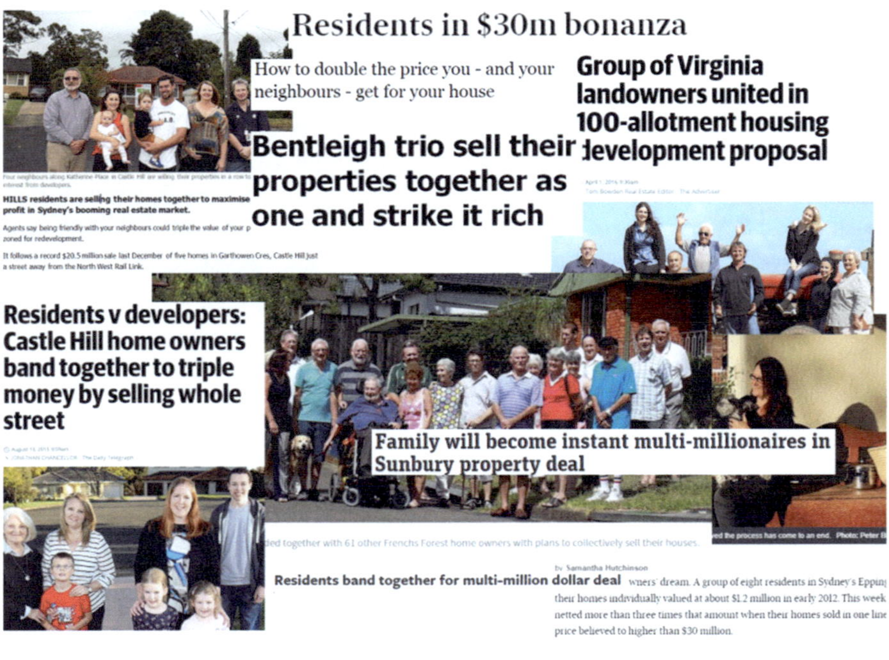

Fig. 9.9 Citizen-led site consolidation in the greyfields

property development options (e.g. sell to developer, partner in a private sector redevelopment, form a co-housing co-operative), statutory planning (e.g. understand impact of current zoning provisions). Community engagement studies undertaken for the GtG TA are reported in Glackin and Newton (2015) and Murray et al. (2015).

Urban Planning Regulation There is unlikely to be any significant advantage – beyond increased dwelling yield – from assembling larger residential sites in the greyfields *if* current planning and building regulations continue to dictate the nature of the redevelopment that can occur. Current regulations remain wedded, for the most part, to managing the micro-level impacts of knock-down rebuild projects (e.g. in Victoria, via ResCode). They do not embrace the innovations that can accompany precinct-scale regeneration – in dwelling design, density, distributed technologies, shared spaces, alternative and reduced on-site parking, green infrastructure, etc.

The manner in which land uses are zoned represents a powerful city-shaping instrument. The most recent residential zonings for Melbourne released in mid-2014 as part of the latest metropolitan strategy plan (*Plan Melbourne 2014*; DTPLI 2014) have been used by a majority of municipalities to 'lock up' any potential for significant intensification of housing by allocating over 90% of existing properties to the zones where redevelopment is restricted either to 'single dwellings and some dual occupancies' (Neighbourhood Residential Zone) or 'single dwellings and some medium density' (General Residential Zone). This represents an attempt by a significant number of municipalities to 'game' the zoning process, introducing inconsistency and a lack of accountability by local government in responding to a metropolitan strategy attempting to plan more sustainably for an entire city (Newton and Glackin 2017). This has not gone unnoticed – from the press (e.g. Donegan 2015) or from the state government (DELWP 2016), who have announced a review of residential zoning for Melbourne.

Plan Melbourne Refresh (2015) represents an attempt by the current (Labor) state government to expose a number of planning policies and strategies that had been proposed by the Ministerial Advisory Committee (MAC) to the previous (liberal) government, but had been totally ignored in *Plan Melbourne 2014*. Drawing on the latest MAC (2015) report, *Plan Melbourne Refresh* (DELWP 2015) contains several recommendations central to urban infill, including a 70:30 target for new housing supply (70% in the established suburbs). Also included was an explicit recommendation related to greyfield precinct renewal (see Fig. 9.10), which is a direct policy response to the *GtG* TA. This endorsement seeks to explore the possibility of a new zone or overlay explicitly developed to support precincts as an alternative to lot-by-lot (KDR) redevelopment. Studies undertaken for the *GtG* TA that explore innovative planning responses for precinct-scale redevelopment are reported in Newton et al. (2011) and Newton and Glackin (2017).

Construction, Delivery and Affordability Part of the *GtG* TA challenge has also been to explore prospects for delivering more affordable medium-density housing in the established middle suburbs of Australian cities. Housing affordability is a major issue in Australia's largest cities (as identified in earlier section on 'landscape')

Greyfield renewal

OPTIONS FOR DISCUSSION

41 Introduce a policy statement in **Plan Melbourne 2016** supporting greyfield renewal and investigate planning scheme mechanisms to achieve coordinated and sustainable renewal of established suburbs.

The MAC (2015) report recommends including greyfield renewal precincts in **Plan Melbourne 2014** as locations for greater housing diversity and density (Recommendation 21).

Greyfields are residential areas where the building stock is near or ending its physical life and land values make redevelopment attractive. Melbourne has many residential areas like this that are typically subject to uncoordinated small-scale infill development which delivers a suboptimal outcome. DELWP is a key sponsor of a project which is currently working with a number of local governments in Victoria, Western Australia and New Zealand and has created tools to help identify, consult on and implement greyfield renewal precincts.

Greyfield areas are a vision of precinct planned infill that provides local government and the community with a framework to better direct and achieve more sustainable outcomes from small scale cumulative change in residential areas.

While greyfield renewal is a new and untested concept, over time with appropriate policy support this may change. The concept of renewing industrial areas (brownfield renewal) is only 25 years old but is now well understood and accepted. **Plan Melbourne 2016** can help initiate this form of innovation which will become more feasible and attractive as the city's housing stock ages.

Fig. 9.10 Policy statement proposing Greyfield Precinct Renewal for Plan Melbourne 2016 (Source: DELWP 2015 Plan Melbourne Refresh Discussion Paper, Government of Victoria, p.57)

and features strongly in *Plan Melbourne Refresh* as a planning and development challenge. The high percentage of households aged 55+ in underoccupied, ageing detached housing is due in part to a minimal retained profit after the sale of a greyfield detached dwelling and purchase of a (newer) medium-density dwelling in the same locality. The high cost of construction of medium-density housing projects in the middle suburbs (30% higher than new detached housing in the same region on a per m^2 basis; Newton et al. 2011) is also a contributor to this.

Prefabricated modular construction, while operating at a low base in Australia, and currently associated with high-rise apartment development, was identified by Newton et al. (2011) as having the potential to significantly influence the mode, material selection and cost of residential construction in the future. Medium-density housing sits between the traditional, non-unionised, trade-based building contractors traditionally associated with detached housing construction and the unionised, high-rise commercial and residential building companies more closely aligned to a manufacturing mode of production. A new 'medium-density development model' for the design and assembly of low- to mid-rise buildings for greyfield precincts that can cost-effectively optimise the delivery of this product would represent an area of innovation yet to emerge beyond concept stage (Mejias 2015).

9.3.3 *Stage 3: Pilot Neighbourhoods*

This stage of *GtG* TA is in its infancy. One intervention to date involves greyfield *public housing* (where state government is the landowner of approximately 5% of total stock), much of which has tended to have been built historically in clusters. An AHURI-funded study of precinct-scale public housing regeneration in two Melbourne municipalities (Murray et al. 2015) has demonstrated how these assets can act as an important *catalyst* for public-private *activation* of a neighbourhood (reconfiguring spaces and adding amenity as well as adding to housing supply at different price points) – achieving outcomes well beyond those traditionally associated with renovation or redevelopment on a lot-by-lot basis. The second intervention involves precinct-scale *private housing* renewal in the City of Maroondah. Community engagement activities have begun at the 'Town Hall' level, working with municipal officers to advertise changes to housing policy and to signal the intensions of local government to implement greyfield precincts as a viable alternative to lot-by-lot housing redevelopment in selected locations (City of Maroondah 2016). This will then shift to 'Kitchen Table' engagements, where community members and landowners are engaged directly, in a codesign process, to jointly assemble (and possibly design and develop) regeneration precincts.

9.4 Conclusion

The jury is currently out in relation to the future of greyfield precinct renewal (GPR). The state government in Victoria has recognised GPR as a prospective new policy for inclusion in *Plan Melbourne 2016*, following its appearance as one of the recommended policy options in *Plan Melbourne Refresh 2015*. Public submissions were sought on these recommendations from October to December 2015. In May 2016 the *Summary of Submissions Report* (DELWP 2016) was published at the same time as the contents of all 373 submissions were made available online (http://www.planmelbourne.vic.gov.au/plan-melbourne-refresh/plan-melbourne-refresh-submissions/refresh-submissions). They confirm that new planning-led policies designed to make 21st cities more sustainable, liveable, inclusive, resilient and productive – in the face of a mounting 'landscape' of challenges – will likely be disruptive to residents in the established, low-density, inner- and middle-ring suburbs of Australian cities. GPR represents a niche planning instrument and development model capable of assisting in the delivery of current metropolitan planning policies related to urban infill and more compact, sustainable cities that are currently lagging. GPR was also designed to enable better alignment between the (top-down) higher-level strategic planning objectives of state planning agencies for the major capital cities with the (bottom-up) local area planning and development approval objectives of municipal government authorities and their resident populations.

Having established the *concept and logic* of greyfield precinct retrofit/regeneration (Newton 2010), the next step was to initiate a substantive *transitions project* capable of developing and trialling the innovations necessary for an implementation and mainstreaming (i.e. scaling up and accelerating) of GPR as a new planning instrument of government and development model for the property industry. What has been presented in this chapter is an example of engaged scholarship and an illustration of how to systematically work towards significant policy change. Planning literature is replete with theories, concepts and frameworks pertaining to urban development and change processes. They range from writings on circuits of financial investment in a capitalist world that impact built environments to the role of built environment sustainability in a rapidly urbanising planet with finite ecological boundaries, to public choice theories centred on the self-interests of both residents and politicians as they play out in local development decision-making or the role of institutions in shaping policy, where structures, procedures and power dynamics among the major agents in a city's regime exert significant path-dependent influences. They constitute the literature on which transition theory (which includes MLF) has been formulated.

As far as frameworks capable of guiding applied planning research orientated towards urban change, the phronetic planning schema of Flyvbjerg (2002) poses key questions, such as: where is urban planning heading, who gains and who loses, is the development desirable and what should be done?; and Kingdon's (2011) policy entrepreneur model suggests that successful transitions require an integration of problem streams (priorities), policy streams (alternatives, solutions) and political streams (decision-makers). These are but two of the many candidates capable of shaping inquiry. However, over the last decade, *transition theories* (such as the descriptive MLP) and *Transition Management methodologies* (such as those that outline stages, objectives, processes and participants for a transition arena intervention) have proven to be highly prospective organising frameworks for guiding the *Greening the Greyfields* (GPR) project since its inception in 2010. The frameworks as applied here have provided clarity in successfully conveying complex concepts and proposals to several funding agencies as well as focusing and staging the multiple research activities involved in delivering the project's multiple outputs.

It is clear that we are at a critical juncture in relation to accelerating a sustainability transition for our cities and in particular the greyfields (refer to Fig. 9.3). The established, greyfield suburbs represent a principal opportunity for urban regeneration; and citizen-led greyfield precinct renewal has a major role to play *if* implemented in *Plan Melbourne 2016* policy, along with other key innovative instruments outlined in this chapter. It will be a test of state (and local) government's capacity to overcome the barriers that clearly exist internally within their institutions as well as externally to the introduction of more disruptive but transformational change. The prize for Melbourne – and other major Australian capital cities – would be the prospect of becoming a highly liveable *and* sustainable city.

9.5 Postscript

Since this chapter was written, 'greyfields' are now featured in Australia State of the Environment: Built Environment report, published by the Commonwealth Government (Canberra, 2017), and 'greyfield redevelopment' is now Policy 2.2.4 in the metropolitan planning strategy – Plan Melbourne 2017–2050 – released in March 2017: a 7-year transition from concept to policy.

Acknowledgements AHURI and the CRC for Spatial Information have provided funding for this research since its inception; and Peter Newman, Shane Murray, Stephen Glackin and Roman Trubka have been significant collaborators on this journey. The author has appreciated several exchanges on this project in Melbourne and Rotterdam with Derk Loorbach.

References

Birkeland J (2014) Resilient and sustainable buildings. In: Pearson L, Newton P, Roberts P (eds) Resilient sustainable cities. Routledge, London

City of Maroondah (2016) Community engagement report: developing a new housing strategy for Maroondah. Maroondah City Council, Melbourne

Crawford R, Fuller R (2011) 'Energy and greenhouse gas emissions implications of alternative housing types for Australia' in State of Australian Cities National Conference 2011 Proceedings, Melbourne

CRC for Water Sensitive Cities (2016) Ideas for fishermans bend. http://watersensitivecities.org.au/new-publication-ideas-for-fishermans-bend/

DELWP (2015) Plan Melbourne refresh discussion Paper. Department of Environmental Land Water and Planning, Melbourne

DELWP (2016) Summary of submissions report. Department of Environmental Land Water and Planning, Melbourne

DIRD (2015) State of Australian cities 2014–2015. Department of Infrastructure and Regional Development, Canberra

Donegan P (2015) The selfishness that's tearing Melbourne apart. The Age, 29 Apr 2015, p 45

DTPLI (2014) Plan Melbourne: metropolitan planning strategy Department of Transport Planning and Local Infrastructure. Victorian Government, Melbourne

Flyvbjerg B (2002) Bringing power to planning research one researcher's praxis story. J Plann Educ Res 21:353–366

Geels F (2011) The multi-level perspective on sustainability transitions: response to seven criticisms. Environ Innov Soc Trans 1:24–40

Geels F, Schot J (2010) The dynamics of transitions: a socio-technical perspective. In: Grin J, Rotmans J, Schot J, Geels F, Loorbach D (eds) Transitions to sustainable development: new directions in the study of long term transformative change. Routledge, New York

Girardet H (2015) Creating regenerative cities. Routledge, London

Glackin S, Newton P (2015) Engaging the Greyfields: community engagement and co-design in residential redevelopment of public housing. In: Legacy C, Leshinsky R (eds) Instruments of planning: tensions and challenges for delivering more equitable and sustainable cities. Routledge, London

Glackin S, Newton P (2016) Assessing the capacity for urban infill in Australian cities. Paper presented at future housing: global cities and regional problems, Melbourne, 9–10 June 2016

Gleeson B, Dodson J, Spiller M (2012) Governance, metropolitan planning and city-building: the case for reform. In: Tomlinson R (ed) The unintended city. CSIRO Publishing, Melbourne

Grin J, Rotmans J, Schot J, Geels F, Loorbach D (eds) (2010) Transitions to sustainable development: new directions in the study of long term transformative change. Routledge, New York

IA (2015) Population estimates and Projections. Infrastructure Australia, Sydney

Kelly JF (2011) The housing we'd choose. The Grattan Institute, Melbourne

Kusher C (2015) Dwelling approvals fall to the second highest month in February, CoreLogic RP Data. Available: http://blog.corelogic.com.au/2015/04/dwelling-approvals-fall-to-the-second-highest-month-on-record-in-february/. Accessed 29 Apr 2015

Loorbach D (2010) Transition management for sustainable development: a prescriptive, complexity-based governance framework. Governance 23(1):161–183

MAC (2015) Plan Melbourne 2015 review, report by the Ministerial Advisory Committee, Melbourne

Mejias L (2015) Solving the housing crisis in San Francisco with factory-built housing technology and regulatory reform. MIT Centre for Real Estate, Boston

Murray S, Bertram N, Khor LA, Rowe D, Meyer B, Newton P, Glackin S, Alves T, McGauran R (2015) Processes for developing affordable and sustainable medium density housing models for Greyfield precincts, AHURI final report no. 236. Australian Housing and Urban Research Institute, Melbourne

Neilson L (2008) The "Building Better Cities" program 1991–96: a nation-building initiative of the Commonwealth Government. In: Butcher J (ed) Australia under construction: nation-building past, present and future. ANU E Press, Canberra

Newton P (2010) Beyond greenfields and greyfields: the challenge of regenerating Australia's greyfield suburbs. Built Environ 36(1):81–104

Newton PW (2012) Liveable and sustainable? Socio-technical challenges for 21st century cities. J Urban Technol 19:81–102

Newton P, Bai X (2008) Transitioning to sustainable urban development. In: Newton PW (ed) Transitions: pathways to sustainable urban development in Australia. Springer, Dordrecht

Newton P, Doherty P (2014) The challenges to urban sustainability and resilience. In: Pearson L et al (eds) Resilient sustainable cities. Routledge, London

Newton P, Glackin S (2013) Using geo-spatial technologies as stakeholder engagement tools in urban planning and development. Built Environ 39:480–508

Newton P, Glackin S (2014) Understanding infill: towards new policy and practice for urban regeneration in the established suburbs of Australia's cities. Urban Policy Res 32:121–143

Newton P, Glackin S (2015) Regenerating cities: creating the opportunity for greyfield precinct infill development. In: Legacy C, Leshinsky R (eds) Instruments of planning: tensions and challenges for delivering more equitable and sustainable cities. Routledge, London

Newton P, Glackin S (2017) Greyfield regeneration: a precinct approach for urban renewal in the established suburbs of Australia's cities. In: Ruming K (ed) Urban regeneration and Australian cities: policies, processes and projects of contemporary urban change. Ashgate, London

Newton P, Newman P (2013) The geography of solar Photovoltaics (PV) and a new low carbon urban transition theory. Sustainability 5:2537–2556

Newton PW et al (2011) Towards a new development model for housing regeneration in greyfield residential precincts, Final report no. 171. Australian Housing and Urban Research Institute, Melbourne. Available: http://www.ahuri.edu.au/publications/download/50593_fr. Accessed 29 Apr 2015

Newton P, Plume J, Marchant D, Mitchell J, Ngo DT (2017a) Precinct information modelling: a new digital platform for integrated design, assessment and management of the built environment. In: Sanchez AX, Hampson KD, London G (eds) Integrating information across the Built environment industry. Routledge, London

Newton P, Taylor M, Newman P, Stanley J, Rissel C, Giles-Corti B, Zito R (2017b) Decarbonising suburban mobility. In: Dai H (ed) Low carbon mobility for future cities: principles and applications. IET, Stevenage

NHSC (2013) Housing supply and affordability issues 2012–13. National Housing Supply Council, Canberra

NSW Government (2011) State environmental planning policy (urban renewal). Department of Planning and Environment, Sydney

Randolph B, Tice A (2015) Suburbanising disadvantage in Australian cities: socio-spatial change in an era of neo-liberalism. J Urban Affairs 36:384–399

Shrinkthatfootprint (2016) How big is a house? Average house size by country. http://shrinkthatfootprint.com/how-big-is-a-house#IlMQkUvLVGvgpouk.99. Accessed 1 June 2016

The Economist (2015) Global house prices. Location, location, location. http://www.economist.com/blogs/dailychart/2011/11/global-house-prices. Accessed 16 Apr 2015

Trubka R, Newman P, Bilsborough D (2010) Assessing the costs of alternative development paths in Australian cities. Curtin University Sustainability Policy Institute, Perth

Trubka R, Glackin S, Lade O, Pettit C (2016) A web-based 3D visualisation and assessment system for urban precinct scenario modelling. ISPRS J Photogramm Remote Sens 117:175–186

Waller V, Blackall L, Newton P (2017) Composting as everyday alchemy – producing compost from food waste in 21st century urban environments. In: Crocker R, Chiveralls K (eds) Reuse in an accelerated world: mining the past to reshape the future. Routledge, London

Watt I (2014) Australia's infrastructure challenge. In: Stanley J, Roux A (eds) Infrastructure for 21st century Australian cities. ADC Forum, Sydney

Wood G, Ong R (2015) The facts on Australian housing affordability. The conversation, 12 June 2015

Chapter 10
The 'Transition Town' Movement as a Model for Urban Transformation

Samuel Alexander and Jonathan Rutherford

Abstract In contexts where governments are failing to deal adequately with urban development challenges from the 'top down', it becomes ever more important to look towards niche grassroot innovations as the key to urban renewal 'from below'. This chapter provides an analysis of one such niche, the nascent Transition Town Movement ('TTM'), which provides one of the more well-known social movements to emerge during the last decade. The fundamental aims of the TTM are to respond to overlapping social, economic, and environmental crises by decarbonising and relocalising the economy through a community-led model of change. With respect to sociotechnical transitions literature, this movement can be understood to be privileging the *social* over the *technical*. Resilience building by urban communities in the face of regime breakdown or deterioration is an important transitions challenge to take seriously, and that is the focal perspective of this chapter. Through a visioning exercise, we explore the role the TTM could play in 'retrofitting the suburbs' as a means of strategically managing foreseeable crises in the incumbent regime, with the aim of turning those crises into opportunities for urban or suburban renewal.

Keywords Transition Town Movement • Grassroot innovation • Resilience • Limits to growth • Social movements

10.1 Introduction

Theories of urban development typically look to governments, local and national, to take the lead in transforming urban landscapes to promote sustainability and wellbeing. This is especially so when the problems requiring a coordinated response – such

S. Alexander (✉)
Melbourne Sustainable Society Institute, The University of Melbourne, Melbourne, VIC, Australia
e-mail: samuelalexander42@gmail.com

J. Rutherford
Simplicity Institute, Melbourne, VIC, Australia
e-mail: johnnyrutherford2@gmail.com

T. Moore et al. (eds.), *Urban Sustainability Transitions*, Theory and Practice of Urban Sustainability Transitions, DOI 10.1007/978-981-10-4792-3_10

as climate change – are deep, urgent, and often 'wicked'. Nevertheless, in many parts of the world today, including Australia, recent and current government policies provide little hope that the range of structural changes necessary to create more sustainable, low-carbon cities will emerge from the 'top down'. Despite paying lip service to sustainability issues, most political actors still operate firmly within an outdated growth paradigm where new roads, new coal mines, or fracking for oil and gas are touted as solutions to urban transport and energy problems, and too often we see cities continuing to eat away at their surrounding greenways with conventional, sprawling, poorly designed housing developments. Business as usual more or less prevails.

So far, as governments are failing us, it becomes ever more important to look towards 'niche' grassroot movements as the key to urban renewal 'from below' (Seyfang and Haxeltine 2012; Smith 2007). In the sociotechnical transitions literature (e.g. Kemp et al. 1998), the term niche often refers to a 'protected space' that a government provides to a promising social, economic, or technological innovation that has the potential to refine or even replace an incumbent system or 'regime' (Geels 2002). But a niche can also refer to innovative activities or practices that lie outside and yet challenge a regime, without being a space protected by the state, as such. Marc Wolfram (2017) argues that cities themselves provide protected spaces in the sense that individuals and communities are able to articulate and enact diverse 'alternative ontologies' and 'spatial imaginaries' of sociotechnical change (Longhurst 2015) because cities fundamentally enable the manifestation of diversity. Furthermore, Gill Seyfang and Adrian Smith (2007: 585) use the term niche to include grassroot initiatives, which they define as 'networks of activists and organizations generating novel bottom-up solutions for sustainable development'. These networks may challenge a regime even if governmental protection or assistance is not provided. Grassroot innovations within urban contexts provide a range of niches that Wolfram (2017) notes has gone 'largely unexplored' within the sustainability transitions literature.

This chapter provides an analysis of one such niche, the nascent transition town movement (hereafter 'TTM'), which provides one of the more well-known social movements to emerge during the last decade in response to overlapping energy, environmental, and economic crises (Hopkins 2009, 2011; Aiken 2012; Barry and Quilley 2008; North and Cato 2012). Whereas the more established ecovillage movement has generally sought to escape the urban context to establish experiments in alternative living, the TTM, motivated by similar concerns, tends to accept the challenge of transforming urban life from *within* the urban boundary. As argued in this chapter, this grassroot movement has relevance to urban transitions, generally (Taylor 2012; Pelling and Navarrete 2011), and Australian urban development, in particular, due to the sprawling, suburban nature of most Australian cities and the uninspired state of Australian politics (Bay 2013; Holmgren 2012; Mason and Whitehead 2012). In the absence of progressive political leadership, might the TTM niche (see Seyfang and Haxeltine 2012) or something like it need to play a role creating new urban landscapes? What potential does this movement has for changing the regime? What are the challenges it might face? And could cities ever become interconnected hubs of transition towns at large?

Since its inception in 2005, the TTM has spread to many countries around the world, including Australia, and is gaining increased attention from academics, politicians, and media (Feola and Nunes 2014). Its fundamental aims are to respond to the overlapping challenges of climate change, peak oil, social isolation, and economic instability by decarbonising and relocalising the economy through a community-led model of change based on permaculture design principles (Holmgren 2002). Rather than waiting for governments to lead, communities in this movement are embracing the 'power of just doing stuff' (Hopkins 2013), getting active trying to build the new world within the shell of the old. In doing so, the movement runs counter to the dominant narrative of globalisation, techno-optimism, and economic growth and instead offers a positive, participatory, highly localised but more humble vision of a low-carbon future, as well as an evolving roadmap for getting there through grassroot activism (Wilson 2012). In the words of Tim Jackson, this international movement is 'the most vital social experiment of our time' (see Hopkins 2011; see also, Broto and Bulkeley 2013).

Recently, Australian permaculture theorist and educator, David Holmgren – whose work has significantly shaped the TTM in theory and practice – has called for grassroot movements, such as the TTM, to 'retrofit the suburbs' (Holmgren 2012). Such a process would involve individuals and communities acting locally – with or without government support – to try to radically transform their urban landscapes by thinking creatively about how to make the best of an infrastructure that is often poorly designed from social and environmental perspectives. Defining activities include attempts to localise food production and connect with local farmers, increase home-based economies, relearn the skills of self-sufficiency, practise frugality and voluntary simplicity to reduce consumption, organise sharing and barter schemes beyond the formal economy, take the energy efficiency of their homes and lifestyles into their own hands, as well as attempt to decarbonise energy use not only through household and community-based renewable energy systems but also by minimising energy consumption through behaviour change (e.g. cycling more and driving less). If this movement evolved from niche activity to mainstream consciousness, how far could it transform the urban landscape in Australia and elsewhere? And what challenges should it expect from the existing regime if it really began to scale up?

This chapter will explore these issues, focusing on the promise and challenges of urban and suburban transformation 'from below'. In applying transitions theory, we acknowledge that the precise delineation of niche and regime cannot always be sharply defined and TTM movement raises special challenges here. Following Seyfang and Haxeltine (2012: 386), we maintain flexibility in our use of regime, recognising that participants in the TTM act across and within various sociotechnical regimes (transport, food, energy, housing, etc.) while at the same time being focused on a 'broad societal transition' away from a regime of fossil fuel dependency, growth, and globalisation. The analysis begins, therefore, by outlining that broad societal regime, which provides the necessary context for understanding the TTM's goals of resilience and relocalisation. After providing that groundwork, the chapter sketches a vision what urban life might look like if the TTM scaled up in response to deepening crisis in the regime and concludes by considering some obstacles that lie in way of such a transition.

10.2 The Regime of Growth and Globalisation

Any theory of 'transition' needs to have both a robust understanding of the current state of things as well as some conception (even if it is constantly evolving) of the goal or ideal state that would represent a successful transition. Only then would anyone be in a position to coherently consider questions of transition – that is, strategic questions about how to get from where we are to where we would like to be.

We acknowledge that this framing sits uneasily with the approach adopted by most sociotechnical transitions scholars (e.g. Rotmans et al. 2001), who have shown a reluctance to describe in advance what a sustainable society might look like. While the literature recognises the importance of beginning with a clear analysis of the existing situation (Loorbach 2010), caution is understandably shown with regard to future envisioning, because the details of what sustainability implies in practice can never be known in advance. Thus, we must constantly adjust our theories of transition accordingly, in the face of ever-changing challenges, risks, and opportunities. Be that as it may, if questions of transition are asked having misunderstood the existing situation, or having misconceived the most appropriate responses to existing problems, then there is a very real risk that one's theory of transition, motivated by the best of intentions, may be applied in ways that fail to effectively produce any positive effect or, worse still, may even be counterproductive to one's cause. After all, if our map is poorly drawn and our compass is broken, we are unlikely to arrive at where we need to go.

In order to understand the TTM, therefore, it is critical to understand the way many of its participants assess the global predicament and what they see as the most coherent forms of social, economic, and political action in light of the overlapping crises of our times. To begin with, this means 'thinking globally' about both justice and sustainability, for only then can one know how best to 'act locally'.

There are now 7.4 billion people on Earth, and recent studies suggest we're heading for more than 9.5 billion by mid-century and 11 billion by 2100 (Gerland et al. 2014). This global population, even if it stopped growing today, is placing tremendous burdens on planetary ecosystems. By all range of indicators (climatic, oceans, deforestation, top toil erosion, resource depletion, biodiversity loss, etc.), the global economy is now in gross ecological overshoot, year-by-year degrading the biophysical foundations of life in ways that are unsustainable (Steffan et al. 2015). Needless to say, consumption practices in the most developed regions of the world are by far the most environmentally impactful, although the developing world seems to be following (or being forced onto) the same high-impact, fossil fuel-dependent industrial path taken by the most developed nations. Let us not pretend that all the talk about 'sustainable development' in recent decades has produced sustainable development.

Despite the global economy being in this overgrown state of ecological overshoot, we also know that billions of people on the planet are, by any humane standard, under-consuming. If these people are to raise their living standards to some dignified level of material sufficiency, as they have every right to do, it is likely that this will place further burdens on already overburdened ecosystems. All these and more are radically calling into question the legitimacy of the high-impact forms of urban life that have evolved in the most developed regions of the world.

And yet, despite the fact that humanity is making unsustainable demands on a finite biosphere, all nations on the planet (including or especially the richest nations) are seeking to grow their economies without apparent limit. This is highly problematic, to say the least, because of the close connection between economic growth (in terms of GDP) and rising energy and resource consumption (Weidmann et al. 2015). It is all very well to point to the potential of technology and efficiency improvements to produce 'green growth', but the fact is that as the world gets distracted by such theoretical possibilities (Alexander 2015: Ch. 1), the face of Gaia is vanishing.

Without going further into detail, suffice to say that most in the TTM seems to frame their analysis of the world by this 'limits to growth' perspective (Meadow et al. 2004; Hopkins and Miller 2012; Turner 2014). They conclude that globalising the high-consumption, energy-intensive ways of living prevalent in the developed regions of the world would be ecologically catastrophic and reject the theory that all nations on the planet can grow their economies while sufficiently 'decoupling' economic activity from environmental impact by way of technological advancement and efficiency improvements. The extent of decoupling required is far too great (Alexander 2016). Efficiency without sufficiency is lost.

The TTM is also generally sceptical about the ability of renewable energy to easily replace fossil fuels – especially the 94 million barrels of oil that are consumed every day – from which it follows that the most developed regions of the world will almost certainly need to adapt to an energy-descent future if 100% renewable energy supply is achieved. But given the close connection between energy and economic activity (Ayres and Warr 2009), reduced energy supply implies that those developed, energy-intensive societies, such as Australia, will need to go through a phase of planned economic contraction, with the aim of leaving sufficient 'ecological room' for the poorest nations to provide a dignified standard of living for those currently destitute. This will require the rich nations to create new 'post-growth' or 'degrowth' forms of economy while at the cultural level variously reimagining the good life beyond consumer culture (Alexander 2015). Tinkering around the edges of growth capitalism will not cut it.

Granted, most people – including most sociotechnical transitions scholars – are not ready to accept these deep implications of the global situation, but only by understanding and acknowledging the true extent of the ecological predicament and the limits of technological and market-based solutions can one understand why the

TTM has emerged in the form it has. It is important to bear this 'limits to growth' perspective in mind when evaluating the TTM's vision(s) of urban transformation outlined below, which might otherwise be interpreted as being too radical. Indeed it is radical in many ways, but this is defended on the grounds that it is a response proportionate to the magnitude and urgency of the overlapping crises we face.

10.3 The Transition Town Niche of Resilience and Localisation

Having outlined, very briefly, the way many in the TTM seem to understand the global predicament, we are now in a position to outline the movement's 'theory of change'. The first point to note here is that if growth economics is the primary cause of the global predicament, then any adequate response to current crises will require transitioning 'beyond growth'. In the words often attributed to Albert Einstein, we cannot solve today's problems with the same thinking that caused them.[1]

If it is the case, then, that governments are incapable or unwilling to question the growth paradigm – which seems to be a fair description of most governments around the world today, including Australia – then it follows that governments will shape their activities and policies (including urban development policies) within a growth paradigm that cannot solve the problems they are trying to address. For this reason, the TTM emerged to explore the potential of a grassroot or community-led approach to societal change 'from below'. The TTM movement therefore provides a challenge to the dominant approach within sociotechnical transitions literature which primarily looks to technological and market-based solutions to existing crises (see Shove and Walker 2007). The TTM can be understood to be privileging the *social* over the *technical*, casting doubt on the idea that existing crises can be solved by or within a globalised capitalist economy focused on economic growth.

The rationale for grassroot activity is that 'if we wait for governments, it'll be too late. If we act as individuals, it'll be too little. But if we act as communities, it might be just enough, just in time' (Hopkins 2013: 45). According to some commentators, this approach represents a 'pragmatic turn' (Barry and Quilley 2008: 2) insofar as it focuses on *doing* sustainability here and now. In other words, it is a form of 'DIY politics' (Barry and Quilley 2009: 3), one that does not involve waiting for governments to provide solutions but rather depends on an actively engaged citizenry to drive the change at the local level.

The paradigm shift of the TTM is articulated around notions of 'decarbonisation' and 'relocalisation' of production and consumption. What this means in terms of transforming urban landscapes will be unpacked further below, but the basic

[1] This section draws upon Esther Alloun and Samuel Alexander 'The Transition Town Movement: Questions of Diversity, Power, and Affluence' Simplicity Institute Report 14 g: 1–25.

dynamic is that decarbonisation is necessary and desirable for reasons of climate change and looming peak oil, and given how carbon-intensive global trade is, decarbonisation implies relocalising economic processes as far as possible as well as consciously moving away from high-consumption lifestyles towards ways of living informed by an ethics of frugality, moderation, and sufficiency.

As well as decarbonisation and relocalisation, another central goal of the movement is to build community 'resilience', a term which can be broadly defined as the capacity of a community or society to withstand shocks and the ability to adapt after disturbances (Hopkins 2009: Ch. 3). Notably, within the TTM crisis in the current system is typically presented not as a cause for despair but as a transformational opportunity, a prospective change for the better that should be embraced rather than feared (Hopkins 2011: 45).

When it comes to applying these broad ideas and concerns in practice, the movement's co-founder, Rob Hopkins (2009), outlines a 12-step roadmap that is intended to help communities start, grow, and run a localised 'transition initiative'. These steps involve setting up a steering group, raising awareness about critical issues, developing visible practical projects, organising activities to 're-skill' the community, and formulating an energy descent action plan. These steps, it should be noted, are fairly generic and demonstrate that the TTM does not propose 'prescriptive solutions' (Hopkins 2009: 137) or a 'one-size-fits-all' approach, but rather constitutes an 'open-ended experiment' (Barry 2012: 114) and a broad rethinking of 'how local economies feed, house, and power themselves' (Hopkins 2012: 74–75).

As one of its defining and most admirable features, the TTM aspires to lead by 'practical example' (Hopkins 2011: 73). Hopkins (2011: 146) emphasises that the movement should not be 'just a talking shop' and that 'practical manifestations' of relocalisation and decarbonisation are essential to create momentum. As he notes, 'a transition initiative with dirt under its fingernails will gain credibility' and thereby attract new people (Hopkins 2011: 146). These projects also offer an opportunity for experiential and social learning, connecting or reconnecting with nature, as well as acquiring new skills. This 'great reskilling', as Hopkins (2011: 152–154) calls it, is an essential aspect of resilience building and developing local adaptive capacities.

As a practical matter, food usually appears as an early focal point of transition initiatives, and many initiatives offer training in permaculture and organic gardening and cooking and preserving food (Pinkerton and Hopkins 2009; Bay 2013). Collective initiatives are also put together to encourage local food provisioning, with the aim of 'delinking food and fossil fuels' (Heinberg and Bomford, in Hopkins 2011: 56) and promoting bioregionalism. For example, many transition initiatives try to set up a community garden/allotment or a veggie box scheme, organise an urban farmers' market, as well as organise fruit tree and nut tree planting days, seed banks, and seed swap days. Other transition activities include establishing local currencies and community-based renewable energy projects and organising carpooling schemes, car-free days, educational film nights, bicycle or sewing workshops and cooperatives, and workshops on energy efficiency in the home and workplace (Hopkins 2011, 2013). These are small-scale illustrations of the ambitious attempt to build resilience and decarbonise the economy and ultimately – in the long term – to fundamentally restructure the economy to support relocalisation and better promote social and ecological justice (North 2010).

The Transition Network was founded in 2006 to 'inspire, encourage, connect, support and train communities' on their 'transition' (see Transition Network 2016). It reinforces the idea of self-organisation, as its objective is not to centralise decision-making but to connect diverse initiatives in order to share experiences, knowledge, skills, and ideas on best practice. In this sense, the Transition Network can be seen as creating a protected space for its own niche, in the hope of replicating, upscaling, and translating its practices to broader society (Seyfand and Haxeltine 2012). As of June 2016, the Transition Network had 1258 initiatives registered in over 43 countries (Transition Network 2016a; Henfrey and Kenrick 2015).

In Australia, there are at least 43 transition initiatives (Earthwise Harmony 2016), so the movement has laid down roots. But it remains small, and a literature search suggests that Australian transition towns have yet to have been the subject of scholarly empirical studies.[2] Part of the reason the TTM is so small in Australia could be because the global financial crisis, which deeply impacted most parts of the world economy, more or less passed over the Australian economy. Social movements rarely emerge to transform economies during prosperous times – the social discontent is not there to ignite widespread activism and engagement. Nevertheless, we will argue that there are aspects to the Australian urban landscape that might be particularly suitable for the TTM to flourish, suggesting that this movement deserves attention despite its currently small size.

In any case, an interesting aspect of the TTM in relation to transitions theory is that while the movement indeed aims to change the regime of growth and globalisation, it is nevertheless resigned to the fact that the regime is almost certainly facing an inevitable descent, no matter what, due to environmental limits to growth and the foreseeable decline of fossil fuels in coming years or decades (Mohr et al. 2015). Thus, resilience building in the face of regime breakdown is an important goal to work towards and the perspective we focus on below. We contend that the TTM movement might only come into its own and scale up when crisis in the regime intensifies and provokes local activism in a way that stable and prosperous (albeit deeply unsustainable) times do not. In this way, the TTM arguably inverts the approach of most transitions theory: rather than looking to the regime to support the niche (via 'strategic niche management'), the niche of the TTM is actually an attempt to strategically manage the foreseeable deterioration of the incumbent regime.

---

[2]But see Feola and Nunes (2014) for an international empirical study, including Australia, but in which the specifically Australian data cannot be isolated from other countries.

10.4 Retrofitting the Suburbs for an Energy-Descent Future

Today, most in the TTM are working on figuring out how to make the best of existing urban infrastructure – a task David Holmgren (2012) refers to as ‘retrofitting the suburbs for the energy descent future’. It makes sense to begin the transition process by improving what already exists rather than seeking to create something entirely new. After all, we are hardly going to knock down the suburbs and build things again from scratch in a ‘greener’ way.

A key advantage of the retrofitting strategy, as outlined below, lies in the power and responsibility it places in individuals, households, and communities just to get started reinventing their neighbourhoods, without waiting for government action – action which, in any case, may not be forthcoming. Indeed, even if government action were to be forthcoming, it may be utterly misconceived (e.g. more roads! more sprawl!) and ultimately do more harm than good by locking in further path dependency. Furthermore, in crisis situations, governments may not always have the resources to solve local problems, in which case, building resilience in advance of potential crises is a sensible task for local communities to undertake.

At the personal level, retrofitting the suburbs might involve, for example, taking in boarders by filling an empty room, to prevent further urban sprawl, putting solar panels on the roof, or retrofitting a poorly insulated house, by putting up thick curtains and sealing gaps in windows and doors to increase energy efficiency and minimise energy consumption. Holmgren (2012) and others in the TTM (Hopkins 2013) are also advocates of transforming the household into a place of production again – as it was in centuries or even decades gone by – and not leaving households merely as places of consumption, which they have essentially become in most developed urban contexts today. At the communal level, the retrofitting transition strategy could mean working on establishing projects such as farmers’ markets, sharing networks, community gardens, food swap groups, skillshare workshops, educational film nights, community energy projects, and local currencies or helping on a local wildlife restoration project. Again, the key aims of this broad strategy would be to decarbonise and relocalise economies as far as possible while enriching community engagement through collective action.

These transition activities are currently small in the greater scheme of things and are easily dismissed by critics (with some justification) as of trifling transformative significance. Our immediate task, however, is to consider what this movement might become if ‘retrofitting the suburbs’ came to be embraced and radicalised, not just by a handful of dedicated and informed social activists but by large segments of a local community or suburb. As implied above, we contend that the most likely ‘spark’ for this movement to scale up is a deepening crisis in the global economy, although Seyfang and Haxeltine (2012) also offer suggestions for how the movement could scale up and translate its message in advance of crisis. What is clear is that over the long term, if the TTM is to be of any deep transformative significance, it must go beyond current retrofitting and transition activities (Trainer 2014) and more fundamentally remake suburbs and towns into highly self-sufficient but interconnected urban villages through collective, grassroot action (Trainer 2010, 2016).

As Hopkins (2011: 72–76) emphasises, 'transitioning' is both an inner and outer process. Change is needed not only in the external physical structures, institutions, and organisations upon which societies rest but also in our worldviews, norms, attitudes, and values. In recognition of this, the transition model of change attempts to weave together the power of imagination, visioning, and storytelling, with the practical manifestation of these alternative narratives, through the engagement of the head, the heart, and the hands (Hopkins 2009).

In that spirit, the following section undertakes a 'visioning' exercise that seeks to provoke reflection on what cities might come to look like if, in response to a regime in crisis, the TTM scaled up and succeeded in transforming the urban landscape. The existing literature often examines successful case studies (Roorda et al. 2014) to help guide thinking and action. The value of more creative visioning exercises, such as the following, reside in their potential to help people break free from dominant development and sustainability narratives and scenarios which, we maintain, almost always assume 'green growth' will provide a smooth transition pathway to sustainability without requiring much in the way of lifestyle change. Our underlying argument is that something like what follows may represent 'life within sustainable limits' far more accurately than those mainstream visions of sustainability.

10.4.1 A Vision of 2030: The Urban Landscape Transformed

Let us, then, jump forward to the year 2030 and imagine what urban life might be like if the TTM – or some confluence of similar movements – managed to scale up and transform urban landscapes as crises within the incumbent regime of growth and globalisation intensify. We focus our thought experiment on a generic Australian suburb of mostly single- or double-story bungalows on blocks of land mostly quarter of an acre or smaller. Other contexts would require creative adaptation in different ways but motivated by similar values and concerns.

In this future suburb, most residents have dug up their backyards and developed productive and diverse permaculture gardens in collaboration with their neighbours, partly motivated by very uncertain and often challenging economic times. A social ethic of frugal self-sufficiency is now widely accepted and practised, driven by these hard economic times. A post-consumerist attitude to waste has emerged, drawing on the Depression era slogan: 'use it up, wear it out, make it do, or do without'.

Many fences within the neighbourhoods have been removed, wholly or in part, to establish small urban farms and to increase social interaction. Most backyards also house chickens and/or ducks for eggs and bees for honey and wax, and a few of the larger blocks keep goats for milk and cheese production. Every backyard has a washing line, so electric dryers have become a thing of the past, and water tanks surround houses and sheds to maximise water collection and minimise the need for main supply. Water consumption is also minimised through simple grey water systems, as well as conscientious showering and washing of clothes.

Most suburban households no longer use flush toilets, having moved to composting toilet systems after learning how to create their own systems at community workshops, and this will minimise the need for energy-intensive infrastructure development for sewerage in the future. It is now accepted that a civilised society in an era of water scarcity should not defecate into potable water and should instead close the nutrient cycle by turning human and animal waste into nutrient-rich soil for local food production, especially fruit trees (Jenkins 2005).

In the front yards and in nature strips, a variety of fruit trees, vines, and berries provide some more of the household's food consumption, although since households are not fully self-sufficient, grains and other food stuffs are purchased from or traded at the vibrant local farmers' markets in the weekend. These markets source their food from the bioregion, minimising 'food miles', building local resilience, and leading to almost completely seasonal diets.

Given that many households no longer rely on a private motor vehicle – relying primarily on bikes and public transport instead – some neighbourhoods have been expanding their gardens and urban farms by creating raised garden beds on underutilised parking lots and even on the sides of roads. People are expected to consult with their neighbours when expanding their gardens into public space, but the expansion of urban agriculture has occurred without explicit council approval. Reminiscent of Havana, in Cuba, the new urban landscape could be well described as a permaculture jungle – an 'edible landscape' crammed with long-lived, largely self-maintaining productive plants, especially on the public spaces, parks, footpaths, and the roads that have been dug up. This also reduces the 'urban heat island' effect. Local water catchment systems, such as swales and ponds, allow for maximum retention and use of rainwater.

This culture of urban food production is significant not just because it builds resilience and minimises the need for the imported produce of industrial agriculture; it also connects people with the cycles of nature and their land base in ways that the 'supermarket generations' did not, positively shaping the human psyche in subtle but important ways (see Louv 2011). The abundance of food grown on public land is considered public property, so no one goes hungry. Given that most people eat low-meat, fresh food diets and exercise more through gardening and cycling, community health is better than it has ever been. The obesity epidemic seems to be quickly fading away. In this way, not only are urban landscapes transforming in positive ways, but so too are the urbanites that inhabit them, each shaping the other in dialectical ways.

Nevertheless, many people lost their jobs in the Second Great Depression of 2019–2023, which was sparked by the US fracking boom going bust, causing the price of oil to rise to levels that placed contractionary pressure on the global economy that was already stagnating under unsupportable levels of public and private debt. Many households that once had two incomes now have to get by on one, or even one part-time income. In order to deal with this challenging economic contraction, people are essentially forced to contribute to the 'informal' economy through practices of home production, and co-housing has been on the rise for about a decade. The 'tiny house' movement has emerged as a creative response to eliminate most cases of homelessness.

Garages that used to house cars have been converted into workshops, where people now produce furniture, clothes, alcohol, tools, and other useful products – for themselves or for trade – as well as provide services like bike, shoe, and tool repairs or education. The days of consumer affluence are gone for most people, but there is a silver lining to all this: when communities take on more of these essential productive functions, they become more resilient and cohesive, and the work is generally more fulfilling because it is often creative and necessary to meet important social needs.

As well as reducing dependence on the formal economy, this thriving local economy also minimises the need for travel to places of work, since many people now work from home or only a short bike ride away. In recent years, councils have been put under tremendous social pressure to maximise space available for bike lanes, and this has helped strengthen the shift away from fossil fuel-dependent cars. Increasingly sections of the suburb are being declared 'car-free'. The emergence of the local economy also means that during the day the suburbs are vibrant, densely populated, and safe centres of economic and cultural activity – and are not emptied from 9 am to 6 pm as they were in the recent past. The fact that there are now sophisticated and interconnected sharing networks within and between suburbs also reduces people's dependence on the formal, monetary economy and greatly reduces wasteful consumption. The Internet remains a useful social tool – keeping global cultural and knowledge transfer alive, facilitating sharing, and widely being used for video conferencing to avoid air travel – but it is no longer the centre of social life.

The predominately local economy has also brought with it large social benefits, as neighbours have come to see each other as interconnected parts of highly localised self-provision. After all, which community is richer: the one where everybody owns their own cheap drill in affluent isolation, or the one where the community is connected via a tool shed where a few high-quality tools are shared amongst the neighbourhood? Given that most households are less affluent than previous eras in monetary and material terms, there is far less discretionary income for paid entertainment, so people now spend their leisure time engaging in low-impact creative activity like music or art, and the new forms of local economy that have arisen also provide many avenues for meaningful and fulfilling self-employment as an artisan, often blurring the boundary between work and leisure.

Energy use, particularly in the form of electricity, is used far more frugally than earlier in the twenty-first century – motivated by high prices – but sufficient supplies for moderate use come from local solar thermal, wind turbines, and household solar PV. Many households have biogas digesters in their backyards, providing a modest portion of the energy needed for heating and cooking. Carefully managed woodlots beyond the urban boundary also provide some fuel for heating, and a small amount of corn ethanol is produced to provide some liquid fuel for the most necessary industrial tasks. High energy prices have incentivised greatly reduced energy demand, which obviously reduced the amount of renewable energy infrastructure that was required to meet demand, making the transition more affordable. Things such as cycling, insulating houses, more limited use of heaters and air conditioners, and localising (and reducing) much production have greatly reduced dependence on fossil fuels and global supply chains. Local forests, farms, and animals provide many necessary resources such as timber, clay, bamboo, leather, wool, oils, medicines, wax, mud, and much else.

Over time the existing housing stock will need to be replaced and built with materials sourced as locally as possible and designed for long-term durability and to the highest standards of energy efficiency. More people and communities take part in the construction of their own homes to reduce costs and liberate people from oppressive mortgages. To limit the resources required, as well as limit the spaces needed to heat and cool, new houses are much smaller than was typically the case in developed nations in earlier decades of the twenty-first century, and they are more densely inhabited.

As well as suburban transformation, many people have also been incentivised to leave dense, high-rise urban centres by the prospect of far cheaper land and housing in small towns in rural Australia. Small-town rural Australia is undergoing a revitalisation in ways that resemble the urban transformation, outlined above, which has taken place in many large cities. In fact, the scale of small towns actually makes the transformation just outlined easier to implement and maintain.

The movement away from industrial agriculture has meant that there is a large demand for workers in the increasing number of small organic farms on the urban periphery, and this rural or peri-urban revitalisation is also putting less pressure for the large cities to expand. Given that these self-transformed urban landscapes have proven successful over the last decade or so, *new* urban developments have come to be shaped in their vision, with presumptions of distributed energy systems, urban farms, bike lanes, and local economies now being supported, rather than inhibited, by planning decisions (Sharp 2014). In this way we can come to understand how bold grassroot social movements can, over time, filter upwards and shape top-down politics (Alexander 2013).

When reflecting on this future scenario, the main point is that most if not all of what has just been described could be achieved by local communities in a relatively short period of time, with or without government support. Ideally, these changes would be made in advance of deepening crisis in the regime, as existing transition towns are attempting to do, but scaling up seems difficult (Seyfang and Haxeltine 2012) and unlikely in cultures that have become accustomed to (unsustainable) affluence. Indeed, there is evidence suggesting that the growth of the TTM is slowing (Feola and Him 2016). Nevertheless, we maintain that the TTM remains relevant because it presents a coherent vision for how communities can adapt positively when – as is quite possible and indeed likely – the incumbent regime of growth and globalisation deteriorates over the coming years and decades as it collides evermore with the limits to growth.

10.5 Limits of the Transition Town Movement

There are huge structural and cultural challenges that lie in the way of this niche scaling up to threaten or replace the existing regime. Below we acknowledge three key challenges.

First, individuals and communities live within structures of constraint – infrastructure, economic, political, cultural, etc. – and those structures shape the ways we live our lives (Sanne 2002). One key challenge faced by social movements seeking urban transformation 'from below' is that some of the structural changes needed to help encourage low-impact living (new energy systems, public transport, bike lanes, etc.) are, it would seem, best made by governments.

Be that as it may, in a context of slow or stagnant political action, the question of how local communities can do the best they can *within existing structural constraints* is a strategic question that deserves heightened attention. After all, if the TTM movement is correct in diagnosing the 'growth paradigm' as central to the problems humanity faces, then governments entrenched in that paradigm are unlikely to foster and are more likely to inhibit the kinds of 'post-growth' sustainability transitions required.

In this light, the TTM declares with worrying plausibility that 'if we wait for governments to act, it'll be too late' (Hopkins 2013: 45). So even if one perceives great potential in governments to hasten sustainability transitions, it would be foolish to deny the importance of grassroot drivers of change (Seyfang and Smith 2007). At the very least, such movements can be seen as creating the cultural conditions necessary for 'top-down' action to take root (Alexander 2013). After all, bold top-down action is unlikely to emerge unless there is a political culture that demands it.

Second, there are some niches that the regime may love to see scaled up and others the regime would resist. For example, if governments could support some new renewable energy technology in ways that ultimately led to a swift decarbonisation transition, then such a regime change in energy provision could be welcomed by the incumbent system. But would the existing regime actually want the transition town niche to succeed in ways outlined above? Probably not. After all, the existing politico-economic paradigm is shaped by the desire for continuous economic growth, and we have seen that the TTM is founded upon a 'limits to growth' perspective that rejects the possibility and even the desirability of limitless growth. The point is that a growth-orientated regime will not want a post-growth niche to scale up, so the TTM may well find that scaling up gets harder, not easier.

Finally, for present purposes, there is a reality, just noted, which the TTM must face: it is, in the greater scheme of things, very small. Transition towns arguably get a disproportionate amount of media and scholarly attention because their stories of (limited) success are often uplifting and full of promise. This can give the impression that the movement is bigger than it really is. But when one looks into the lived reality of transition towns, one often learns that most of the initiatives are driven by a very small handful of engaged activists within populations that almost entirely are unengaged with transition activities (Feola and Nunes 2014).

10.6 Conclusion

It is often expected that academic studies conclude with some notes of optimism, highlighting paths forward that will solve the problems that have been under examination. We will resist that convention in order to highlight our contribution to the transitions literature. The logic of our analysis has been quite simple: (1) Governments do not seem to be doing much to facilitate sustainability transitions; therefore, hope arguably resides in the social sphere, which must drive the movements for change. (2) But, promising social movements like the TTM are struggling to scale up and are likely to continue to struggle due to structural lock-in; (3) therefore, a plausible future scenario is one in which the incumbent regime is not replaced voluntarily in a planned and orderly manner but is replaced instead through a series of social, economic, and environmental crises as the incumbent regime breaks down.

Instead of exploring how the regime could manage a niche such as the TTM to avoid such crises, our approach in this chapter was to explore the role the TTM could play in managing a regime that looks destined to deteriorate in the coming years and decades (Turner 2014). So in response to the question 'how might the TTM scale up', we assumed that it may require a deepening crisis of global capitalism, which could provide the incentive to get active in one's community in a way that life does not yet provide for most people living in affluent societies, including Australia. If it turns out that the TTM fails to transform the urban landscape according to its ambitious vision, then *trying* to do so now arguably still remains a good use of time, resources, and energy, because this may help build local resilience as the existing regime grows itself into ever-deepening crises.

Without false optimism, then, our challenge in the coming years may best be understood as managing as positively as we can a regime destined for decline. We have argued that the TTM or something like it provides a coherent pathway for turning that looming crisis into an opportunity for urban renewal.

References

Aiken G (2012) Community transitions to low carbon futures in the transitions towns network (TTN). Geogra Compass 6(2):89–99

Alexander S (2013) The voluntary simplicity movement and the social reconstruction of law: degrowth from the grassroots up. Environ Values 22(2):287–308

Alexander S (2015) Prosperous descent: crisis as opportunity in an age of limits. Simplicity Institute, Melbourne

Alexander S (2016) Policies for a post-growth economy. MSSI issues paper No 6: 1–15
Ayres R, Warr B (2009) The economic growth engine: how energy and work drive material prosperity. Edward Elgar, Cheltenham
Barry J (2012) The politics of actually existing unsustainability. Oxford University Press, Oxford
Barry J, Quilley S (2008) Transition towns: "survival", "resilience" and sustainable communities – outline of a research agenda. Adv in Ecopolit 2:14–37
Barry J, Quilley S (2009) The transition to sustainability: transition towns and sustainable communities. In: Leonard L, Barry J (eds) Ch. 1. The transition to sustainable living and practice. Emerald Group, Bingley
Bay U (2013) Transition town initiatives promoting transformational community change in tackling peak oil and climate change challenges. Aust Soc Work 66(2):171–186
Broto V, Bulkeley H (2013) A survey of urban climate change experiments in 100 cities. Glob Environ Chang 23:92–102
Earthwise Harmony (2016) Transition towns Australia. http://www.earthwiseharmonycom/CONNECT/EH-Transition-Towns-Australiahtml. Accessed 6 June 2016
Feola G, Him MN (2016) The diffusion of the transition network in four European countries. Environ Plann A 48 (11): 2112 – 2115
Feola G, Nunes JR (2014) Success and failure of grassroots innovations for addressing climate change: the case of the TTM. Glob Environ Chang 24:232–250
Geels F (2002) Technical transitions as evolutionary reconfiguration processes: a multi-level perspective and a case study. Res Policy 31:1257–1274
Gerland P et al (2014) World population stabilization unlikely this century. Science 18 September 2014
Henfrey T, Kenrick J (2015) Climate, commons, and hope: the TTM in global perspective. http://climatesecurityagenda.org/wp-content/uploads/2015/11/018138c_TNI_Climate-Commons-Hope.pdf. Accessed 6 June 2015
Holmgren D (2002) Permaculture: principles and pathways beyond sustainability. Holmgren Design Services, Hepburn
Holmgren D (2012) Retrofitting the suburbs for an energy descent future. Simplicity Institute Report *12i*, 1–9
Hopkins R (2009) The transition handbook: creating local sustainable communities beyond oil dependency (Australian and New Zealand edition). Finch Publishing, Lane Cove
Hopkins R (2011) The transition companion: making your community more resilient in uncertain time. Chelsea Green Publishing, White River Junction
Hopkins R (2012) Peak oil and transition towns. Archit Design 82:72–77. doi:10.1002/ad.1432
Hopkins R (2013) The power of just doing stuff: how local action can change the world. UIT/Green Books, Cambridge
Hopkins R, Miller A (2012) Climate after growth: why environmentalists must embrace post-growth economics and community resilience. Post-Carbon Institute Report: 1–28
Jenkins J (2005) The humanure handbook: a guide to composting human manure, 3r edn. Chelsea Green Publishing, White River Junction
Kemp R, Schot J, Hoogma R (1998) Regime shifts to sustainability through processes of niche formation: the approach of strategic niche management. Tech Anal Strat Manag 10(2):175–198
Longhurst N (2015) Towards an 'alternative' geography of innovation: alternative milieu, socio-cognitive protection and sustainability experimentation. Environ Innov Soc Trans 17:183–198
Loorbach D (2010) Transition management for sustainable development: a prescriptive, complexity-based governance framework. Governance 23(1):161–183
Louv R (2011) The nature principle: human restoration and the end of nature-deficit disorder. Algonquin Books, New York
Mason K, Whitehead M (2012) Transition urbanism and the contested politics of ethical place making. Antipode 44(2):493–516
Meadows D, Randers J, Meadows D (2004) Limits to growth: the 30-year update. Chelsea Green Publishing, White River Junction
Mohr S et al (2015) Projection of world fossil fuels by country. Fuels 141:120–135

North P (2010) Eco-localisation as a progressive response to peak oil and climate change – a sympathetic critique. Geoforum 41:585–594

North P, Scott Cato M (2012) A suitable climate for political action? A sympathetic review of the politics of transition. In: Pelling M, Manuel-Navarrete D, Redclift M (eds) Climate change and the crisis of capitalism: a chance to reclaim self society and nature. Routledge, London, pp 99–113

Pelling M, Navarrete DM (2011) From resilience to transformation: the adaptive cycle in two Mexican urban centers. Ecol Soc 16(2):11. http://www.ecologyandsociety.org/vol16/iss2/art11/. Accessed 15 Jan 2013

Pinkerton T, Hopkins R (2009) Local food: how to make it happen in your community. Green Books, Totnes

Roorda C et al (2014) Transition management in the urban context: guidance manual. DRIFT, Erasmus University, Rotterdam

Rotmans J, Kemp R, Asselt M (2001) More evolution than revolution: transition management in public policy. Foresight 3(1):15–31

Sanne C (2002) Willing consumers – or locked-in? Policies for sustainable consumption. Ecol Econ 42:273–287

Seyfang G, Haxeltine A (2012) Growing grassroots innovations: exploring the role of community-based initiatives in governing sustainable energy transitions. Environ Plann C: Gov Policy 30:381–400

Seyfang G, Smith A (2007) Grassroots innovations for sustainable development: towards a new research and policy agenda. Environ Politics 16(4):584–603

Sharp N (2014) The zillion year town. https://archive.org/details/TZYT1. Accessed 2 June 2016

Shove E, Walker G (2007) CAUTION! Transitions ahead: politics, practice and sustainable transition management. Environ Plann A 39:763–770

Smith A (2007) Translating sustainabilities between green niches and socio-technical regimes. Tech Anal Strat Manag 19(4):427–450

Steffan W, Richardson K, Rockstrom, J et al (2015) Planetary boundaries: guiding human development on a change planet. Science 347(6623) (13 February 2015)

Taylor PJ (2012) Transition towns and world cities: towards green networks of cities. Local Environ 17(4):495–508

Trainer T (2010) Transition to a just and sustainable world. Envirobook, Sydney

Trainer T (2014) Transition townspeople, we need to think about transition (Just doing stuff is far from enough!). http://thesimplerway.info/TRANSITIONERS.htm. Accessed 3 June 2016

Trainer T (2016) Remaking settlements: the potential cost reductions enabled by the simpler way. Simplicity Institute Report 15e: 1–39

Transition Network (2016) Official initiatives by number. https://www.transitionnetwork.org/initiatives/by-number. Accessed 6 June 2016

Turner G (2014) Is global collapse imminent? MSSI Research Paper No.4. Melbourne Sustainable Society Institute: 1–21

Weidmann T et al (2015) The material footprint of nations. Proce Natl Acad Sci USA 112(20):6271–6276

Wilson GA (2012) Community resilience, globalization and transitional pathways of decision-making. Geoforum 43:1218–1231

Wolfram A (2017) Grassroots niches in urban contexts: exploring governance innovations for sustainable development in Seoul. Procedia Engineering (in press)

Part IV
Spatial Dimensions of Urban Transitions

Part IV is the last main section in the book and presents three chapters which engage with the spatial dimensions of urban transitions. Chapter 11 explores suburbs as sociotechnical regimes and the impact that lock-in mechanisms have and the historical preconditions that created this outcome. Chapter 12 then uses the example of live/work as a niche innovation within urban planning as a way of analysing the pathways of policy innovation and city development and remaking. Concluding this section of the book, Chap. 13 draws on empirical work to explore mobility challenges in cities and the impacts for equity and sustainability.

Chapter 11
Another Suburban Transition? Responding to Climate Change in the Australian Suburbs

Tony Dalton

Abstract This chapter considers the idea of destabilising the current high-carbon regime and establishing the preconditions for a new sociotechnical regime in Australian suburban cities. It does this in the following four sections. The first section argues that cities can be the site of sociotechnical regimes. In this case, the focus is on the suburbs as a sociotechnical regime within Australian cities. The second section describes the pattern of direct and indirect household energy consumption in large metropolitan cities, which are overwhelmingly suburban cities. This urban/suburban location of high energy-intensive household living is an integral element of the high-carbon sociotechnical regime. The third section argues that the underlying 'lock-in mechanisms' producing and reproducing the suburbs have at times been destabilised and reconfigured. It is important to understand what made the new 'lock-in mechanisms' viable because this can inform strategic thinking about future change. The fourth section draws a set of preconditions from the history of change in 'lock-in mechanisms' that should be considered in the development of transition to low-carbon suburban suburbs. It presents them at three levels – macro, meso and micro – as a means for clarifying the way different types of power is exercised in the making and remaking of energy intensive suburbs. The challenge is how might households live in and remake their cities while they continue to be suburban so that they are more sustainable.

Keywords Suburbs • Energy • Housing • Housing policy • Lock-in • Households • Historical review

11.1 Introduction

Households in Australian cities contribute to greenhouse gas emissions and to global warming and climate change through their use of energy. The most obvious way in which households contribute to greenhouse gas emissions is through their direct use

---

T. Dalton (✉)
Centre for Urban Research, RMIT University, Melbourne, VIC, Australia
e-mail: tony.dalton@rmit.edu.au

T. Moore et al. (eds.), *Urban Sustainability Transitions*, Theory and Practice of Urban Sustainability Transitions, DOI 10.1007/978-981-10-4792-3_11

of energy for services such as space heating and cooling, heating water, cooking, running appliances and lighting. When compared to other developed urbanised countries, Australian households rank somewhere in the middle of the range. The less obvious way that households contribute to greenhouse gas emissions is through other practices associated with living in a dwelling located in a city. They include obtaining and eating food, disposing of waste, travelling to and from work and other destinations, using and disposing of water and the purchase and use of many other goods and services produced outside the home. Through these practices households use what is described as indirect or embodied energy. The research shows that in industrialised countries, including Australia, 'the fraction of indirect requirements is usually on a par with or even greater than the direct energy requirements' (Wiedenhofer et al. 2013).

There has been strong growth in awareness amongst households of direct energy use, how much they use it and how they use it. Sustained increases in electricity and gas prices have been important in prompting households to understand the ways in which they can moderate their use of gas and electricity. For many, they have done this by changing their appliances, improving the thermal efficiency of their dwelling and modifying domestic practices so that they achieve a more efficient use of gas or electricity and lower their energy costs. Also, a significant proportion of households have responded by generating their own energy by installing photo voltaic panels, solar hot water heaters and heat pumps (Australian Energy Market Operator 2016; Bahadori and Nwaoha 2013). Other households have been more constrained, typically low-income households in private rental market, in the way they have been able to respond to increased prices. They have not been able to modify their dwellings, change their appliances or invest in new technologies. Instead many of these households have experienced energy poverty by having to reduce their energy use and experiencing less domestic comfort and amenity (Moore and Dalton 2016).

There is less household awareness about indirect energy use. As Lenzen et al. (2008, p. 91) observe: 'We have often no idea how much energy is needed to produce the things we buy, and whether all of the embodied energy is higher than the direct energy we experience using in our daily lives'. In part this is because energy reporting typically uses industry categories such as transport, services, agriculture and manufacturing. Therefore, it is difficult to discern how much energy has been used to produce services or goods purchased and used by households. This difficulty has led researchers to develop different and competing analytical methods that can produce estimates of how much indirect energy is used by households in cities (Kok et al. 2006). Even though they can only produce estimates, it is nevertheless important to include indirect energy use calculations for two reasons. First, household members are end user consumers for a large proportion of total consumption. Second, the quantum of indirect energy consumed through households is continuing to grow and in high-income countries exceeds direct energy (Lenzen et al. 2008, p. 94; Reindersa et al. 2003).

In summary, households through their direct and indirect use of energy are consuming a significant proportion of total energy output. Further, because direct and indirect energy is overwhelmingly sourced from fossil fuels, households are making a significant contribution to greenhouse gas emissions. In this context, it is important to consider how these living arrangements, which constitute a carbon-intensive sociotechnical regime, might transition to a low-carbon sociotechnical regime. This could be through a mixed strategy of reducing energy use through changing practices and increased energy efficiency and substituting renewable energy for energy generated by burning fossil fuels. However, a transition of this nature will not happen easily. As Geels (2010, p. 395), drawing on Unruh (2000), notes that 'socio-technical transitions to sustainability do not come about easily, because existing energy, transport, housing and agri-food systems are stabilised by lock-in mechanisms that relate to sunk investments, behavioural patterns, vested interests, infrastructure, favourable subsidies and regulations'. For example, it cannot be assumed that current challenges, such as the new politics of energy use, costs and supply that constitute a 'landscape-level pressure' (Geels and Schot 2007) on the existing carbon-intensive suburban city regime, will disrupt the lock-in mechanisms that hold the regime together. It is therefore important to examine more systematically the nature of lock-in mechanisms and how they might be destabilised and the possibility of support growing for new lock-in mechanisms that support a low-carbon suburban sociotechnical regime.

This chapter considers the idea of destabilising the current high-carbon regime and establishing the preconditions for a new sociotechnical regime in Australian suburban cities. It does this in the following four sections. The first section argues that cities can be the site of sociotechnical regimes. In this case the focus is on the suburbs as a sociotechnical regime within Australian cities. The second section describes the pattern of direct and indirect household energy consumption in large metropolitan cities, which are overwhelmingly suburban cities. This urban/suburban location of high energy-intensive household living is an integral element of the high-carbon sociotechnical regime. The third section argues that the underlying 'lock-in mechanisms' producing and reproducing the suburbs have at times been destabilised and reconfigured. It is important to understand what made the new 'lock-in mechanisms' viable because this can inform strategic thinking about future change. The fourth section draws a set of preconditions using three levels of analysis – macro, meso and micro – from the history of change in 'lock-in mechanisms' that should be recognised by those promoting strategies aimed at a transition to low-carbon suburban suburbs.

The objective of this paper is to contribute to the debate about how Australian cities can become more sustainable. It is not a paper aimed at contributing to the polarised for-and-against the suburbs debate that has been a feature of Australian urban studies. It recognises that Australian cities are suburban cities – the suburban form is a defining characteristic, and because new housing per annum forms such a small proportion of total stock, a move away from the suburban form is slow. In this context, the challenge is how might households live in and remake their cities while they continue to be suburban so that they are more sustainable. Davison (2006, p. 202) suggests that this future sustainability rests on the capacity of actors to bring together the social and technical through 'public debate over the rough ground of practical deliberation, as distinct from technical calculation, where scientific rationality and wise judgement are joined together'.

11.2 Suburban Housing as a Sociotechnical Regime

The suburbs of Australian cities, overwhelmingly comprised of stand-alone suburban dwellings, constitute a sociotechnical regime. Rip and Kemp (1998, p. 338) provide a starting point for understanding the nature of a sociotechnical regime as a:

> rule-set or grammar embedded in a complex of engineering practices, production process technologies, product characteristics, skills and procedures, ways of handling relevant artefacts and persons, ways of defining problems – all of them embedded in institutions and infrastructures.

Applied to suburban housing, the sociotechnical regime can be described in the following way.

First, there are the distinct rule-sets that guide land developers who have engineered the garden city-type road layouts of curved streets, shopping centres and cul-de-sacs, as the suburbs have moved inexorably further out. Each house lot is connected invisibly to the network infrastructures of water, sewerage, telecommunications, electricity and gas. Second, there are the products, suburban houses, chosen by households from builder catalogues and built using an elaborate trade-based on-site subcontract system for an agreed price (Dalton et al. 2013b). Purchaser households usually fund the purchase of their house, either new or existing, by 'taking out' a mortgage with a bank. In recent times Green Star regulations have required new dwellings to meet modest higher energy and water efficiency standards.

The houses, located on lots and connected to the network infrastructures, have distinct characteristics (King 2016, p. 172).

> We can see a house as a complex machine. It has lots of parts that we expect always to be working when we want them, such as heating systems, flushing toilets, showers, cookers, lighting and electricity. Much of the moving parts are hidden from view, behind walls, and so we lose sight of the complexity. At least we do until something stops working. We then come to realise just how complex the dwelling is, and we find that we need experts to repair and maintain it.

Second, the repair, maintenance and renovation of housing, or the remaking of housing, is another significant feature of the suburban housing sociotechnical regime. Measured in terms of gross fixed capital formation, it is almost as significant as investment in new housing (Dalton et al. 2013a, p. 22). This is similar to other suburban housing societies with mature housing systems (Horne and Dalton 2014, p. 3446). Associated with this level of investment are highly mobile and dispersed trade-based businesses spread out across suburban cities engaged by households to add extra floor area, maintain, repair, renovate and redecorate existing dwellings.

Within this sector an eco-renovator business niche has developed in response to homeowner households seeking to improve the energy and water efficiency of their dwelling (ibid).

Third, the development of a suburban culture has accompanied the making and remaking of the suburbs. The suburbs and the houses are redolent with meaning. This became evident as the suburban form of development was institutionalised in the late nineteenth century (G. Davison 1995). Of course the culture has continued to change as households have continued to make and remake their homes through a multitude of practices associated with daily life. More recently there is a sense in which this meaning has become imbricated with ideas of about wealth and superior levels of comfort, convenience and consumption. Allon (2008, p. 207) observes:

> The home became the meaning of life: buying a home, buying a bigger and better home (and often more than just one), renovating the home we already had, became the driving pursuits for most. We took our rising incomes, our growing wealth, and poured it into our homes.

In the UK Smith (2008, p. 521), who describes a similar housing market phenomenon, suggests that owned homes are now best understood as a 'hybrid of money, materials and meanings'.

The suburbs in Australian cities constitute a sociotechnical system that has been created through the development and application of distinctive technologies of land development, housing construction, reinvestment and financing. Institutions frame the legal and financial arrangements for their production, and various workforces get on with the production and reproduction of suburban space and dwellings. The households that live in these dwellings, in collaboration with the producers and reproducers, generate meanings about property ownership and life in the suburbs. The challenge is that the households living in these suburbs are using more direct and indirect energy which is contributing to greenhouse gas emissions and global warming.

11.3 Energy Consumption

Household direct and indirect energy consumption in cities has been increasing, associated with the 'elevated energy and general resource intensity of urban economies and lifestyles' based on cheap fossil fuels (Droege 2008, p. 2). However, energy consumption across cities is not uniform, and this increasing consumption, although ubiquitous, varies spatially within cities. Also, average direct and indirect energy consumption differs between cities with similar histories and level of economic development. In other words, there is evidence of structural factors at work which result in a distinctive pattern of both direct and indirect household energy use. The research literature that compares and models energy use in the urban context provides insights into the complexities of household energy use and the structural factors that drive growth. This section focuses on the following factors that help to understand the pattern: household income and expenditure, household size and composition, ageing population, residential built form, household appliance growth and use and urban transport. It should be recognised however that different analytical modelling approaches have distinct strengths and weaknesses (Swan and Ugursal 2009) and there is a continuing debate about the efficacy of the different methodologies.

Household income and expenditures are the most significant driver of energy use by household members (Estiri 2015; Lenzen et al. 2008; Wiedenhofer et al. 2013). However, it is expenditure that gives the best indication of energy use because it is a measure, based on household expenditure surveys, of what households actually consume. Direct energy use for space heating and cooling, lighting, water heating and powering appliances increases with household income. However, as income increases the cost of direct energy use also declines as a proportion of total household expenditure. In a sense expenditure on direct energy levels off. However, increasing expenditure on indirect energy embedded in goods and services purchased by households maintains a tighter relationship with increasing household expenditure. In other words consumption does not level off and as income rises households tend to use more indirect energy and contribute more to greenhouse gas emissions than do lower-income households.

Household size and composition influence both direct and indirect energy consumptions. Smaller households for both forms of energy consumption have higher per capita consumption, and concomitantly larger households have lower per capita consumption. The more people who occupy a dwelling means the more people there are to share the direct energy services and goods and services containing embodied energy. In this context it is important to note the growth in the proportion of single-person households in Australian cities and the forecast growth in this proportion (Australian Bureau of Statistics 2015). However, there is a caveat to this general rule of thumb which relates to the presence of children in the household. For example, Gibson et al. (2013) discuss the resource implications of children's use of nappies in the first 2–3 years of life before toilet training. And Dodson and Snipe (2008) note the energy use of families with young children living in outer suburbs poorly serviced by public transport. Further, between 1971 and 2003, there is the observed shift from walking to journeys by car in children's mode of travel to and from school (van der Ploeg et al. 2008; Whitzman 2013).

The increasing proportion of aged person households is increasing per capita direct and indirect household energy use. This occurs because children leave home and because an increasing proportion of people are entering into old age as single people through earlier divorce and separation and through the death of a partner. This phenomenon is sometimes referred to as 'under-occupation', especially in the context of multi-bedroom suburban housing where children have left home. In addition, there are also changing patterns of energy use associated with ageing. They include more time spent in the home which can result in increased use of energy for heating and cooling. However, for some aged, because of low incomes, they experience 'energy poverty' and restrict their use of energy and lower their comfort levels (Chester 2013). For other higher-income aged households, energy use can increase through greater use of recreational use of air travel (Hamza and Gilroy 2011) and local car use (Gibson et al. 2013). It is clear that there are many changes underway, especially in the context of 'baby boomer' retirement, but there is no systematic accounting of energy use by retirees (Gibson et al. 2013).

The residential built form can influence energy use. Typically a distinction is made between detached and multiunit dwelling structures. Lee and Lee (2014, p. 535) in their review of the literature conclude that households in 'multifamily housing units, characterized by shared walls and typically smaller floor space, consume less energy for space heating, cooling, and all other purposes than do households in detached single-family homes, when controlling for the age of housing structures as a proxy of construction technology'. In their modelling of energy use in Sydney, Wiedenhofer et al. (2013, p. 704) confirm this relationship when they find that higher population density 'due to less per capita floor space in higher density residential areas, where flats and semi-detached houses are dominant, which also tend to be more energy-efficient than separate houses'. However, as Perkins et al. (2009, p. 394) show, comparing inner-area apartments, inner-terrace housing and suburban detached housing in Adelaide, the design, use and management of higher density buildings 'can render central city apartment living less greenhouse friendly'. There is also preliminary evidence that the five-star residential building energy efficiency standard is moderating direct energy use, especially for space heating (Ambrose et al. 2013).

Growth in the number and use of appliances in households has had a profound long-term effect on increasing household direct and indirect energy use. Electrical energy use associated with household appliances in developed countries grew by 57% from 1990 to 2005 (IEA 2008, p. 48). This growth was strongly associated with growth in the use of large appliances. More recently it has been associated with the growth in the household use of small miscellaneous appliances such as personal computers, mobile phones, personal audio equipment and other home electronics (IEA 2008, p. 48). Australian households have participated in the growth in the number of appliances, which is forecast to continue for the next 20 years. However, direct energy consumption associated with appliance use has declined largely due to the increasing energy efficiency of appliances (Australian Energy Market Operator 2016). Whether this decline continues will depend on whether energy standard efficiency regime keeps up with appliance growth and innovation (Saddler 2016). However, the indirect or embodied energy in these products is significant. Lenzen et al. (2008, p. 98), using 1995 data, calculate that the average amount of indirect energy in the manufacture of a typical appliance is approximately ten times the direct energy used per annum. Total energy consumption associated with household appliance acquisition and use is therefore continuing to increase.

Growth in the use of private motor vehicles has been the key feature in urban transport in contemporary urban development resulting in the steady increase in energy use and greenhouse gas emissions. Lenzen et al. (2008, p. 99) describes the trend. 'In absolute terms household transport by private car represents the majority of transport energy use ... linked principally to income, both car ownership and mileage per vehicle have increased'. Australia has been a stand out country, third behind the US and Canada, in its reliance on car based transport (Kenworthy 2008, p. 224). However, this dependence in Australian cities is not uniform. Instead it is closely related to residential location, and car use and energy use increase with distance from the city centre (Dodson and Snipe 2008; Lenzen et al. 2008). The pattern of dependence has also been changing, and there is now a case that in many cities including Australian cities, there is evidence of a decline in the share of private car travel in metropolitan cities and an increase in the public transport share. It is suggested that there has been a peaking in private car use (Kenworthy 2013; Newman et al. 2013).

Two conclusions can be drawn from this brief review of six drivers that produce the distinctive pattern of both direct and indirect household energy uses in cities.

First, the pattern of direct and embodied energy use and the resulting CO_2 emissions are deeply embedded in the daily practices of households. It is shaped by a complex mix of decisions about expenditures, choices about household size and composition such as the choices about the number of children, the choices of older people about staying or moving as household size reduces, housing stock design and condition, continuing renewal and uptake of new appliances and decisions about how to move around the city. Also the resulting pattern of direct and indirect energy use across Australian city is distinct. Wiedenhofer et al. (2013, p. 704) summarise the outcome for metropolitan Sydney:

> In the more urban and wealthier districts, indirect and total energy requirements are highest, which translates into higher consumption of all kinds of goods and services. At the same time direct energy use, for example petrol or electricity, is lowest for urban households, compared to average suburban and rural households.

Second, we can observe that household members have agency, some more than others, in the way they spend money, form and dissolve households, choose a dwelling, buy appliances and decide how to travel. However, if we borrow Marx's insight and apply it to the minutiae of daily life in cities, rather than the possibilities for revolution, we can observe that household members do these things in context. They 'make their own history, but they do not make it as they please; they do not make it under self-selected circumstances, but under circumstances existing already, given and transmitted from the past'. In Australian suburban cities, household members participate in an already well-established energy-intensive suburban sociotechnical regime. Further, it is a sociotechnical regime that enjoys broad support, is understood to be in equilibrium and does not require disruption.

11.4 Changing Mechanisms for Producing and Reproducing the Suburbs

Nevertheless, there are critiques, ideas and niches organised around alternative practices that are based on the idea that the energy intensity of the suburbs must be lowered if decarbonisation targets are to be met. This is the context for asking how best to think about a transition in the energy-intensive suburban sociotechnical regime. One way to respond to this question is to respond with normative sensible ideas about what should and could feasibly happen next. However, this approach runs the risk that the preconditions for the implementation of these ideas are missing. Debates about city futures are redolent with good ideas about how to design and redesign cities that remain on the wayside, particularly ideas proposed by urban professionals working within well-established professional paradigms.

Instead this section identifies key episodes in the history of the suburbs when 'lock-in mechanisms' were reconfigured. It does so on the grounds that behind the seeming permanence of suburban landscapes and built form, there have been sociotechnical transitions. The history of the suburbs is not one of the uninterrupted linear progress. Instead it there have been periods of stability and disruption in the social and economic relations that enable households to purchase, rent and live in the suburbs. These changing relations are less visible than the permanence of the houses and network infrastructures that connect them to energy, communications, water and waste disposal and transport systems. Knowledge of past transitions in the social and economic relations have guided suburban development might assist in understanding how a future transition might be considered.

This approach responds to the challenge issued by Avelino and Rotmans (2009, p. 546) that there is a 'need to incorporate a consistent conceptualisation of power', They base this challenge on the finding that transition studies are about change, but there is often little explicit consideration of the power relations underpinning sociotechnical regimes and their destabilisation and subsequent realignment. They find that there is too much faith put in the idea that transitions can be 'managed'. Instead argue for analysing power in relation to actors and the way they mobilise and exercise power, the types of power and the level of power they exercise.

Four episodes that stand out in the history of the suburbs are establishing the owner-occupied garden city suburb, the owner-occupied suburb as an object in the regulation of the Australian finance system, public housing suburbs in support of industrial development and financial deregulation and free market suburbs. Figure 11.1 presents these four episodes graphically against the growth of the housing stock and the change in the tenure mix of this stock. A brief analysis of each of the episodes is presented with a particular focus on the way in which the actor groups define the issues and mobilise. In each episode the role of the state through its agency arrangements is central to understanding the way the issues are problematised and the policy responses.

11.4.1 The Owner-Occupied Garden City Suburb

In the late nineteenth century metropolitan urban development in Australia was imperilled through the collapse in the terms of trade, a depression and the collapse of the colonial financial systems. This collapse included the building societies

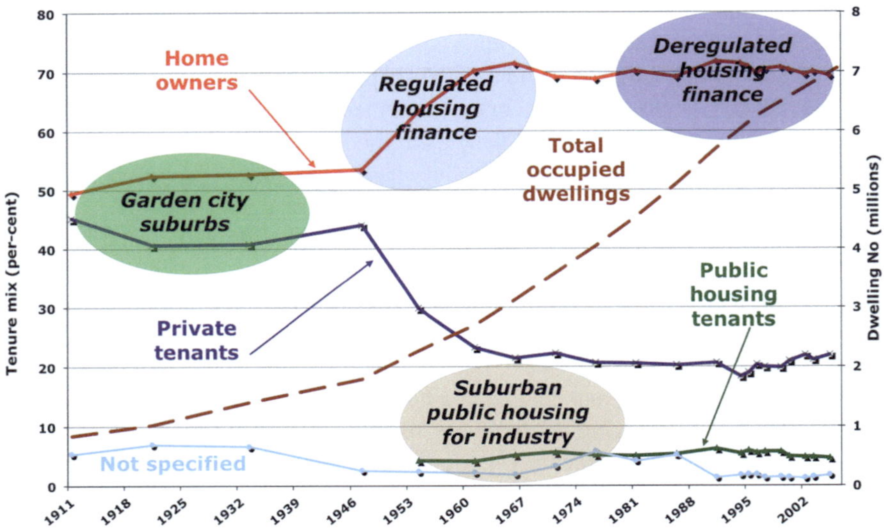

Fig. 11.1 Four key episodes shaping the suburbs in Australia

that channelled domestic savings into residential development. Along with this collapse went their capacity to teach 'a fully fledged suburban ideology of home ownership' embodying the virtues of security, stability' and appropriate gender relations (G. Davison 1981, p. 175). This is the context within which the state became a key player in the elaboration of a suburban sociotechnical regime that guided the growth of the suburbs in the early decades of the twentieth century.

Three key institutional developments underpinned the development of the regime: suburbs as the setting for model households, new government agencies supporting suburban expansion and housing industry associations and alliances.

The increasing presence and organisation of social liberals became a force for developing and proselytising the determinist idea that the urban physical environment influenced the character of city dwellers and the health of society. They constituted a new social group who were found in the ranks of the organised professions of doctors, lawyers, architects, town planners, philanthropists, clergy, labour leaders and journalists. Their objective was to 'rationalise the domestic world: to extend the principles of science and instrumental reason to the operation of the household and to the management of personal relationships' (Reiger 1985, p. 3). A particular focus was inner-city slum residents and promoting new suburbs as an alternative through inquiries and royal commissions with new dwelling designs, garden city movement-inspired street layouts and homeownership.

The development of new state capacities was another institutional project. This began by establishing authorities with the capacity to provide network infrastructures for sewerage, water, gas, electricity and light and heavy rail public transport necessary for land development and suburban expansion. The development of independent and stable state government-owned savings banks with a branch presence in the suburbs accompanied the development of network infrastructure authorities. The banks developed a capacity for drawing in household savings, providing long-term credit foncier home loans to households. In Victoria the State Savings Bank began designing and supervising the construction of 'Bank Homes' for purchasers (Murray and White 1992, p. 212). After WW I the federal government established the War Service Homes Commission and collaborated with the state banks to expand the supply of long-term mortgages to returned servicemen while imbricating patriotic and suburban ideologies.

Accompanying the development of the suburbs came the formation of associations and alliances of builders that sought to order relations between producers, subcontractors, employees and customers of suburban housing. The Master Builders Association has its origins in this period. In addition, the building material suppliers formed a succession of associations around the production of timber, plaster, clay building products and metal products. Employees also came together to form craft unions and federated with trades and labour councils. They focussed primarily on employment conditions but also became involved in broader housing industry and housing provision issues. By 1931 in Victoria, the industry groups including employer organisations, unions, suppliers, real estate agents and professional groups came together to form the Building Industry Congress and became involved in broader debates, principally about slum reform (Howe 1988, p. 21).

11.4.2 Suburban Housing and Regulated Finance

Urban housing production crashed during the depression, and by the end of WWII, housing shortages in the context of a baby boom and a growing immigration programme became an issue. This led to major growth in the provision of suburban housing. During the 1950s an average of 75,000 dwellings were built each year. During the 1960s the annual rate of construction increased and reached a high point of 150,000 dwellings in 1974. In the three decades, 1947–1976, the stock of dwellings increased from 1.9 to 4.1 million dwellings, a 121% increase. At the same time the rate of owner occupation increased from 53 to 68% of households. It was a period of considerable metropolitan city expansion centred on growth in the suburban owner-occupied housing stock.

Two key institutional developments underpinned the expanded capacity of the regime: the development of a national economic policy capacity and the strengthening capacity of housing industry interests to speak for the industry and influence policy.

Within national government the core executive economic policy agencies, especially Treasury, the Reserve Bank of Australia and the Prime Ministers Department, developed the capacity to manage an economic growth model and forge a social consensus, inspired by Keynesian economic theory, that relied on an active central state (Catley 1996; Smyth 1994; Watts 1987). A key element was establishing and managing a 'protected circuit of capital' to fund the expansion of owner-occupied housing, based on quantitative controls and regulated interest rates. The government-owned Commonwealth Bank and the state government-owned banks operated this system by retailing mortgages to households who bought suburban houses or used the funds to buy land and build new houses on the fringe. On occasions these arrangements were overlaid with additional assistance programmes such as first homeowner grants and subsidised interest rates. Central government *dirigisme* driven by economic policy agencies led the growth of Australian post-war suburbs which confirmed Australia as an owner-occupier society.

An accompanying institutional development was a greater role for producer interests sometimes in cooperation with real estate interests, trade unions and community welfare and civic organisations. They spoke on behalf of households who had experienced the privations of housing shortages caused by the cessation of housing construction during the Depression and WWII and subsequent rapid population growth. Their demands were for mechanisms that expanded the protected circuit of capital and increased the share of national savings being invested in housing. They became an influential force with both sides of politics and were successful through several mobilisations in achieving increases in the capital available for investment in owner-occupied housing and going a long way towards meeting the post-war housing shortage. The capacity of producer interests to influence policy continues as a feature of national housing policymaking (Gurran and Phibbs 2015).

11.4.3 Suburban Public Housing

State governments began large-scale construction programmes producing suburban public rental housing in the decades following WWII in 1945. Central government supported these programmes by allocating significant amounts of capital through fixed interest rate long-term loans. By 1955 the State Housing Authorities (SHAs), the agencies established to develop and run the public housing programmes, produced 18,000 dwellings in that year. The private sector completions in this same year were 64,000. A decade later in 1966, they were still producing 17,000 dwellings. The principal way the SHAs produced these houses was to acquire the land and then to contract out the construction to private builders many of whom were building for sale to owner-occupiers. In this sense government was integrally involved in building the post-WWII capacity of suburban land developers and house builders.

Two key institutional developments underpinned the expanded development of public housing in the suburbs: the development of government authorities able to design, construct and manage suburban rental housing and a coalition of interests that supported the establishment of a new tenure in the suburbs of capital and regional cities.

SHAs became large semi-independent authorities with a capacity to design, construct and manage large-scale housing programmes (Jones 1972). In the case of the Victorian Housing Commission, this included developing a capacity for technical and construction system innovation (Dalton 1988). Most public housing was linked to industrial development through suburban housing estate development adjacent to the new industries. In this way suburban public rental housing supported industries such as power generation, car manufacturing, steel making and ship building. At the local level the SHAs developed an administrative capacity to collect rent, allocate housing to households on waiting lists, transfer households to other houses and maintain properties. SHAs continue to provide housing although their role is diminished. Public housing stock numbers have not grown following cuts in funding. And following several decades of economic restructuring, their primary role is to provide housing to households who are excluded from the labour market and very reliant on income security programmes.

The public housing programme only developed because it enjoyed a broad coalition of support. A starting point for this support came from the enterprises who were looking for a well-housed proximate workforce. For example, the power generator in Victoria was the State Electricity Commission, another large government instrumentality. The steel maker was BHP and the car manufacturers were GMH and Ford. Local government authorities were also strong supporters because they were keen to attract employers and new residents to their municipality. There was also support from the architecture and engineering professions who supported some SHA efforts, especially the Victorian Housing Commission, to shift the mode of housing construction from on-site craft production to factory-based industrial production. Civic organisations who fiercely supported the idea of homeownership also lent their support when some SHAs began to sell public housing to tenants.

11.4.4 Suburban Housing and Finance Deregulation

Central government closed down the protected circuit of capital for owner-occupied housing in the mid-1980s. They did this by removing interest rate controls on new mortgages and quantitative controls. Policy proposals that proposed keeping the protected circuit but modernising it were rejected. This policy shift followed the broader ascendency of neo-liberal thinking. This shift was tellingly signalled in the opening doctrinal statement of the Committee of Inquiry into the Australian Financial System (Campbell Inquiry) (1981, p. 1) which stated 'the discipline of the market remains the most economically efficient basis for allocating funds and resources, from the viewpoint of the community as a whole'. The housing market since financial deregulation has been transformed (Yates 2011). Amongst these changes finance availability and the ability for households to borrow have increased which in turn has contributed to long-term sustained house price increases, increasing household debt and declining housing affordability (Berry and Dalton 2004).

Two key institutional outcomes have been associated with this transformation: the ascendancy within central government of the agencies that drew on neo-liberal thinking and the weakness of social liberals to put together a unified set of ideas and build them into new institutions.

The core executive agencies, the Treasury, the Reserve Bank of Australia and the Prime Ministers Department, were central in designing the arrangements for deregulating the finance system. Housing issues were viewed quite differently. Economic policy advisers viewed housing as just one commodity amongst many and subjected it, in the same way, to the new economic calculus, which did not include century-old suburban owner-occupation cultural values. This culture was subordinated to a new economic policy framework which held that a deregulated finance system was a precondition for a more competitive Australian economy in a rapidly changing global economy. Establishing institutional frameworks, even with reduced and more transparent subsidies, that could be used to channel investment into housing supply in line with urban and social policy objectives was defeated. Urban housing outcomes are not a key performance indicator in the contemporary economic policy calculus.

Another outcome has been the failure of social liberal forces to establish sufficiently robust institutional capacities necessary for influencing housing policy and programme outcome. In the debates around the future of the finance system and policy responses, they struggled to come to a consensus. One group sought to defend the status quo. Another group agreed with elements of the neo-liberal critique of housing policy and programmes and at the same time tried to devise new financing mechanisms such as indexed bonds for an expanded social housing sector that responded to the housing needs of a wider income group. These proposals were opposed by the core executive agencies and were defeated (Dalton 1999). The inability of social liberal forces to institutionalise a new set of arrangements able to allocate capital to a broad-based social housing sector continues.

11.4.5 A Synthesis

Each episode resulted in reconfiguration in the social and economic relations of the growing suburbs resulting in significant change in the way the suburbs were understood as a place to live, forms of ownership and how they should be financed. The accounts of these episodes responded to the Avelino and Rotmans (2009, p. 546) challenge that there is a 'need to incorporate a consistent conceptualisation of power', which included a focus on the actor groups that participate in the exercise of power. This is further clarified by identifying common themes in the way actors exercised power in these episodes. The themes are presented by ordering them in terms of macro, meso and micro levels. The macro draws out the themes that were significant at the broader societal level; the meso is the middle ground where the issues play out and the actor groups are visible; and the micro is the households who populate the suburbs.

At the macro level a feature of each of the four episodes was the support that came from broadly accepted ideas. When state governments began their programmes of support for garden city layouts and commenced their mortgage lending programmes, social liberal ideas, following the political and economic turmoil of the 1890s, provided support for reorganised interventionist state governments that could lead the creation of new settler society cities. Keynesian ideas shaped policy thinking in the wake of the 1930s depression and WWII. Subjugation of the power of financial institutions and policy that directed the allocation of capital was simply accepted, as was the idea that governments had a role in establishing a public rental housing system that supported industrial development. By the 1980s the neo-liberal ideas had become ascendant, and Keynesians were on the defensive. The discipline of the market in allocating funds was now the guiding doctrine guiding the flow of funds in the economy.

The second macro feature in each of the four episodes was the institutional capacity of the state to translate the ideas into policy and programmes. In Australia this capacity was established by the last decade of the nineteenth century when 'bureaucratic autonomy had been attained and a pattern of state regulation of civil society was established' (Deacon 1989, p. 127). This capacity was evident in the state governments when they established state banks, urban infrastructure authorities and SHAs. It was also evident in the development of federal government creation of agencies, in particular the Treasury, the Reserve Bank of Australia and the Prime Ministers Department. These agencies had the capacity to translate Keynesian ideas into policy and programmes and later repeat the process of refashioning the financial system drawing on neo-liberal ideas.

The third macro feature evident in each of the four episodes is the producer associations representing the interests generated through the housing, land and finance industries. This began with the development of the owner-occupied garden city period and has continued as these associations have grown and developed. Associations such as the Housing Industry Association, Master Builders Association, Property Council of Australia, Urban Development Institute of Australia and Australian Bankers Association are well established with state and national secretariats that service their members and are constantly presenting policy positions to government (Gurran and Phibbs 2015). Their progenitors were active during the Keynesian period, and by the time of financial deregulation, they were active in their current form.

The meso level is the arena where the state agencies come together with the associations (Atkinson and Coleman 1989). It is in this arena that the associations and state agencies form what can be called policy networks. Broadly these are more or less stable and regularised interactions between these actor groups. They are evident in processes such as industry conferences, advisory committees, committees of inquiry, consultations, delegations and lobbying. Their levels of activity can rise and fall depending on the issues. Their role through the various processes is to identify and discursively define 'problems' and negotiate on policy change drawing on shared assumptions about what is feasible. Using the distinction between reactive and anticipatory policy networks suggested by Atkinson and Coleman (1989, p. 60), the housing policy network can be described as largely reactive.

At the micro level in this schema are the households living in the suburbs. They are the households that over time flow through the stock of suburban houses largely as purchasers and less so as renters. As they live in the suburbs, these households create social and economic relationships full of meaning associated with their practices of daily living. However, as Smith (2008) reminds us, an important dimension of these relationships is the economic relationship that the household has with the 'housing market' and the meaning that the household ascribes to it. In the UK, the context for Smith's research notes the most recent period is characterised by 'financial marketisation' and has led to an 'encounter' between the 'grammars of living' for owner-occupation (which increasingly hinge around the asset value of housing) and the 'rules of the game' negotiated by households as they engage in the meanings, materiality and multiple values associated with homeownership. This is also the case in Australia where house price asset inflation and the relationship of housing wealth to consumption and long-term security are pronounced.

11.5 Conclusion

The suburbs of Australian cities can be understood as constituting a sociotechnical regime. They are produced by different types of firms who combine to use distinctive technologies to convert rural land to urban land and build houses. Financial institutions provide the finance necessary for households to purchase the houses. Another set of firms are then involved in the maintenance, repair renovation and refurbishment of housing. The households who live in the new and the existing housing stock through their practices are then active participants in continually ascribing meaning about living in the suburbs. It is possible to look back and see how this culture has changed over time. In the current context of declining housing affordability, increasing house prices and housing reinvestment, the meanings are increasingly linked to the financial value and status of houses.

The problem is that energy use and greenhouse gas emissions associated with suburban living have been increasing. The suburbs have steadily been becoming more energy intensive and contributing to climate change. The challenge is to consider how suburbs might transition from high carbon to low carbon. This can be approached at two levels. First, there is direct use of energy by households using electricity, gas and private car use. There has been some moderation in the average use of electricity and gas use as prices have risen over the last decade. Second, there is indirect energy use by suburban households. This form of energy use comes through household purchase and use of services and goods produced elsewhere. This form of energy use has been increasing, principally through increased ownership and use of appliances. There is a distinct spatial patterning of direct and indirect energy use. This is the ubiquitous high consumption context that challenges the modest aspirational efforts of participants in the Transition Towns Movement discussed in Chap. 10 of this book.

The prospects for a transition are explored by looking back at the history of the suburbs. Beneath the materiality of the suburbs, there have been at least four distinct episodes in the 'lock-in mechanisms' used to produce and reproduce the suburbs. These episodes were reviewed by examining the ideas that underpinned the search for a new model and the key actor groups that were involved. By examining the ideas and the actors, there is the possibility of understanding how power is exercised in sociotechnical change processes within the suburbs. A review of these episodes shows how important prevailing hegemonic ideas are to being able to define what the problem is that needs to be solved. This review also shows how important it is for the state through its policy and administrative agencies to have the capacity to lead policy development and implementation. This history of state involvement in past transitions challenges the argument, such as that made in Chap. 10 in support of the Transition Towns Movement, that decarbonisation of the suburbs might be achieved largely without state policy change and resources. Finally, this review showed how influential the producers of the suburbs have been in participating in the state-mediated policy networks that produce negotiated policy solutions. This history of producer interests suggests just how important the success of the greyfield precinct regeneration idea, discussed in Chap. 9, depends on policy networks able to institutionalise the necessary support and means for remaking the older carbon-intensive suburbs.

Based on this analysis the prospects for a sociotechnical transition of the suburbs from high carbon to low carbon are not propitious. In relation to direct energy use, there has, as Moloney and Horne (2015, p. 2451) describe, been a modest history of policy and programme initiatives aimed at initiating a low-carbon transition in Melbourne. However, the conclusion they draw is that 'low carbon transitioning is unstable, localised and transitory'. In relation to indirect energy use in the suburbs, there is no broader focus on a transition leading to significantly reduced indirect energy use through reduced consumption. Instead there are the initiatives of households and groups of households who are organising from the bottom up such as the Transition Towns Movement. This is apparent in aspirations of urban farmers' markets, suburban food production and composting, appliance sharing, exchange schemes and so on. There is a type of reliance on the original producers of goods and services continuing to reengineer their supply chains in ways that further increase their energy productivity.

Yes, there are continuing 'landscape-level pressures' on business as usual in the carbon-intensive suburbs such as declining housing affordability and homelessness, increasing congestion and journey to work times, increasing energy prices and supply uncertainties, social and economic infrastructure deficits and lack of renewal of existing infrastructure and new climate change-related vulnerabilities to events such as droughts, bushfires, cyclonic winds, flooding and heatwaves. However, they are not as yet generating the type of coherent policy processes able to seriously challenge the 'lock-in mechanisms' that continue to produce and reproduce the current high-carbon regime and prefigure preconditions for a new low-carbon sociotechnical suburban regime.

References

Allon F (2008) Renovation nation. University of New South Wales Press, Sydney

Ambrose MD, James M, Law A, Osman P, White S (2013) The evaluation of the 5-star energy efficiency standard for residential buildings. Retrieved from Canberra

Atkinson MM, Coleman WD (1989) Strong states and weak states: sectoral policy networks in advanced capitalist economies. Br J Polit Sci 19(1):47–67

Australian Bureau of Statistics (2015) Household and family projections, Australia, 2011 to 2036. Retrieved from Canberra

Australian Energy Market Operator (2016) National electricity forecasting report. Australian Energy Market Operator Ltd.

Avelino F, Rotmans J (2009) Power in transition an interdisciplinary framework to study power in relation to structural change. Eur J Soc Theory 12(4):543–569

Bahadori A, Nwaoha C (2013) A review on solar energy utilisation in Australia. Renew Sust Energ Rev 18:1–5

Berry M, Dalton T (2004) Housing prices and policy dilemmas: a peculiarly Australian problem? Urban Policy Res 22(1):69–91

Catley B (1996) Globalising Australian capitalism. Cambridge University Press, Melbourne

Chester L (2013) The impacts and consequences for low-income Australian households of rising engergy prices. Department of Political Economy, University of Sydney, Sydney

Committee of Inquiry into the Australian Financial System (Campbell Inquiry) (1981) Final report. Australian Government Publishing Service, Canberra

Dalton T (1988) Architects, engineers and rent collectors: an organisational history of the commission. In: Howe R (ed) New houses for old: fifty years of public housing in Victoria. Ministry of Housing and Construction, Melbourne

Dalton T (1999) Making housing policy in Australia: home ownership and the disengagement of the state. Unpublished PhD thesis, RMIT University, Melbourne

Dalton T, Chettri P, Corcoran J, Groenhart L, Horne R (2013a) Understanding the patterns, characteristics and trends in the housing sector labour force. Positioning Paper 142. Australian Housing and Urban Research Institute, Melbourne

Dalton T, Hurley J, Gharaie E, Wakefield R, Horne R (2013b) Australian suburban house building: industry organisation, practices and constraints. AHURI Final Report No 213. Australian Housing and Research Institute Melbourne.

Davison G (1981) The rise and fall of marvellous Melbourne. Melbourne University Press, Melbourne

Davison G (1995) Australia: the first suburban nation? J Urban Hist 22(1):40–74

Davison A (2006) Stuck in a Cul-de-Sac? Suburban history and urban sustainability in Australia. Urban Policy Res 24(2):201–216

Deacon D (1989) Managing gender: the state, the new middle class and women workers 1830–1930. Oxford University Press, Melbourne

Dodson J, Snipe N (2008) Shocking the suburbs: urban location, homeownership and oil vulnerability in the Australian City. Hous Stud 23(3):377–401. doi:10.1080/02673030802015619

Droege P (2008) Urban energy transition: an introduction. In: Droege P (ed) Urban energy transition: from fossil fuels to renewable power. Elsevier, Burlington

Estiri H (2015) The indirect role of households in shaping US residential energy demand patterns. Energ Policy 86:585–594

Geels F (2010) Ontologies, socio-technical transitions (to sustainability), and the multi-level perspective. Res Policy 39:495–510

Geels F, Schot J (2007) Typology of sociotechnical transition pathways. Res Policy 36(3):399–417

Gibson C, Farbotko C, Gill N (2013) Household sustainability: challenges and dilemmas in everyday life. Edward Elgar Publishing, Cheltenham

Gurran N, Phibbs P (2015) Are governments really interested in fixing the housing problem? Policy capture and busy work in Australia. Hous Stud 30(5):711–729

Hamza N, Gilroy R (2011) The challenge to UK energy policy: an ageing population perspective on energy saving measures and consumption. Energ Policy 39(2):782–789

Horne R, Dalton T (2014) Transition to low carbon? An analysis of socio-technical change in housing renovation. Urban Stud 51(16):3445–3458

Howe R (1988) Reform and social responsibility: the establishment of the housing Commission. In: Howe R (ed) New houses for old: fifty years of public housing in Victoria. Ministry of Housing and Construction, Melbourne, pp 1938–1988

IEA (2008) Worldwide trends in energy use and efficiency: key insights from IEA indicator analysis. Energy indicators. International Energy Agency, Paris

Jones M (1972) Housing and poverty in Australia. Melbourne University Press, Melbourne

Kenworthy J (2008) Energy use and CO_2 production in the urban passenger transport systems of 84 international cities: findings and policy implications. In: Droege P (ed) Urban energy transition: from fossil fuels to renewable power. Elsevier, London

Kenworthy J (2013) Decoupling urban car use and metropolitan GDP growth. World Transp Policy Prac 19(4):7–21

King P (2016) The principles of housing. Routledge, London

Kok R, Benders R, Moll H (2006) Measuring the environmental load of household consumption using some methods based on input–output energy analysis: a comparison of methods and a discussion of results. Energ Policy 34:2744–2761

Lee S, Lee B (2014) The influence of urban form on GHG emissions in the U.S. household sector. Energ Policy 68:534–549

Lenzen M, Wood R, Foran B (2008) Direct versus embodied energy – the need for urban lifestyle transitions. In: Droege P (ed) Urban energy transition: from fossil fuels to renewable power. Elsevier, Burlington

Moloney S, Horne R (2015) Low carbon urban transitioning: from local experimentation to urban transformation? Sustainability 7:2437–2453

Moore T, Dalton T (2016) Structuring housing provision for sustainability. In: Horne R, Fien J, Beza B, Nelson A (eds) Sustainability citizenship: living and working sustainably in our cities. Routledge, London

Murray R, White K (1992) A bank for the people: a history of the state bank of Victoria. Hargreen Publishing, Melbourne

Newman P, Kenworthy J, Glazebrook G (2013) Peak car use and the rise of global rail: why this is happening and what it means for large and small cities. J Transp Technol 3(4):272–287

Perkins A, Hamnett S, Pullen S, Zito R, Trebilcock D (2009) Transport, housing and urban form: the life cycle energy consumption and emissions of city Centre apartments compared with suburban dwellings. Urban Policy Res 27(4):377–396. doi:10.1080/08111140903308859

Reiger K (1985) The disenchantment of the home: modernising the Australian family. Oxford University Press, Melbourne

Reindersa A, Vringerb K, Blok K (2003) The direct and indirect energy requirement of households in the European Union. Energ Policy 31:139–153

Rip A, Kemp R (1998) Technological change. In: Rayner S, Malone E (eds) Human choices and climate change, vol 2. Battelle, Columbus, pp 327–399

Saddler H (2016) Australia's carbon emissions and electricity demand are growing: here's why. The Conversation, April 18th, https://theconversation.com

Smith SJ (2008) Owner-occupation: at home with a hybrid of money and materials. Environ Plann A 40:520–535

Smyth P (1994) Australian social policy: the Keynesian chapter. University of New South Wales Press, Sydney

Swan LG, Ugursal VI (2009) Modeling of end-use energy consumption in the residential sector: a review of modeling techniques. Renew Sust Energ Rev 13(8):1819–1835. doi:http://dx.doi.org/10.1016/j.rser.2008.09.033

Unruh GC (2000) Understanding carbon lock-in. Energ Policy 28:817–830

van der Ploeg HP, Merom D, Corpuz G, Bauman AE (2008) Trends in Australian children traveling to school 1971–2003: burning petrol or carbohydrates? Prev Med 46(1):60–62. doi:http://dx.doi.org/10.1016/j.ypmed.2007.06.002

Watts R (1987) The foundations of the National Welfare State. Allen & Unwin, Sydney

Whitzman C (2013) Harnessing the energy of free range children. In: Low N (ed) Transforming urban transport: the ethics, politics and practices of sustainable mobility. Routledge, London

Wiedenhofer D, Lenzen M, Steinberger J (2013) Energy requirements of consumption: urban form, climatic and socio-economic factors, rebounds and their policy implications. Energ Policy 63:696–707

Yates J (2011) Housing in Australia in the 2000s: on the agenda too late? Paper presented at the Reserve Bank of Australia Conference, The Australian Economy in the 2000s, Sydney, 15–16 Aug

Chapter 12
Emerging Theoretical Space: Urban Planning and Sustainability Transitions

Andréanne Doyon

Abstract Live/work refers to combined dwelling and workplace in a single unit or property. Live/work has the potential to contribute positively to a city at the neighbourhood scale from multiple perspectives, including social, economic, and cultural, as well as environmental. Policies for live/work are still far from widespread, and as such, they can be viewed as niche innovations within an urban planning regime. This chapter tracks the trajectory of live/work as a niche intervention from the informal to formal to present day using the multilevel perspective (MLP) and compares the policy adoption process in Vancouver and Melbourne.

Using an embedded multiple-case-study approach with a theoretical replication design, this research explores contrasting results between two cities. In Vancouver, live/work has been integrated into the regime, whereas in Melbourne, it has remained outside the system. To analyse the trajectory of live/work, documentation, interviews, direct observation, and physical artefacts were used as data collection methods. This investigation is concerned with demonstrating how live/work accelerates as a sustainability transition and which governance structures, approaches to planning, or stakeholders influence those processes. It is identified that rigid and top-down governance structures are less flexible and open to change, political approaches to planning are less responsive and adaptive, and strong political actors have the ability to either initiate or inhibit change.

Keywords Policy innovations • Urban transitions • Urban planning • Urban governance

12.1 Introduction

Since the beginning of the twentieth century, single-function buildings have been favoured by zoning laws, building codes, and lending policies, which were supported by legally enforced separated zoning for *work* and *live* uses, resulting in

A. Doyon (✉)
RMIT University, Melbourne, VIC, Australia
e-mail: andreanne.doyon@rmit.edu.au

T. Moore et al. (eds.), *Urban Sustainability Transitions*, Theory and Practice of Urban Sustainability Transitions, DOI 10.1007/978-981-10-4792-3_12

urban monocultures (Alexander et al. 1977; Davis 2012). When the contemporary live/work movement got traction in the 1980s and 1990s, beginning with the artist loft movement, it challenged the current urban planning system. This chapter explores the emergence of live/work as a 'new' land-use and building typology in two cities: Vancouver and Melbourne.

This chapter tracks the trajectory of live/work as a niche intervention from the informal to formal to present day using the multilevel perspective (MLP) and compares the policy adoption process in Vancouver and Melbourne. Particular attention is given to which governance structures, approaches to urban planning, and stakeholders influence the transition process. As a land use, live/work has been used as a tool for urban revitalisation, as a building type it has fostered hubs of creativity and intervention, and both definitions push against the traditional, Euclidean (single-use zoning) planning systems. Live/work has the potential to contribute positively to a city at the neighbourhood scale (Newman and Jennings 2008) from multiple perspectives, including social, economic, and cultural, as well as environmental (Friedman 2012).

This chapter begins by situating this research within three facets of sustainability transitions research: geography of transitions, urban transitions, and the MLP. This is followed by an introduction to the discipline of urban planning, which is useful in setting up the development of the analytical framework. Next, the methodology for the research is presented, including the case studies of Vancouver and Melbourne, and live/work as the niche intervention under investigation. This chapter then discusses the findings from the two case studies separately; afterwards the success factors and challenges are analysed as a cross-case comparison. The intersection or combination of fields of sustainability transitions and urban planning is revisited at the end as an emerging theoretical space.

12.1.1 Responding to the Literature: Geography of Transitions, Urban Transitions, and the Multilevel Perspective

Within the geographical 'space' of sustainability transitions, urban transitions are identified as a geographic category (Hansen and Coenen 2015). Cities are believed to be strategic sites for the support and management of transitions (Truffer and Coenen 2012) as 'urban and regional policies often run ahead of national and supranational regulations in their response to climate change' (Hansen and Coenen 2015, p. 102). However, with sustainability transitions research, Hodson and Marvin (2010, p. 480) raised the issue 'of where cities "fit" within the multilevel perspective and, in particular, where do cities sit within the landscape-regime-niche hierarchy?' They believe that cities can be included within either the regime or niche and conceive of cities as receiving transitions from high national transitions to be implemented or mediated at the local level or as places where transition initiatives develop. Cities and urban regions can act as protected spaces to embed and diffuse niches and offer policies to stimulate certain industries. Gibbs and O'Neill (2014) argue that cities can also span across landscapes, depending on how the boundaries of a study are framed. Other scholars believe that cities and urban regions are 'place bound' and therefore provide a natural analytical unit to study the process of sustainability transitions (Gallopín et al. 2001; Pickett et al. 2003; Smith and Stirling 2010. See also Chaps. 1 and 2).

This chapter proposes using the city as a 'bounded system' and the 'place' for research using a modified MLP. This research, therefore, responds to calls made for a more explicit account of the geography of sustainability transitions (Truffer et al. 2015). In addition, many scholars involved in this emerging theoretical space believe that the established frameworks (such as technological innovation systems (TIS), MLP, strategic niche management (SNM), and Transition Management (TM)) would benefit from a more serious engagement with geography to more effectively understand the influence of space and place and spatial differentiation and uneven development processes (Coenen and Truffer 2012; Murphy 2015; Truffer et al. 2015).

The MLP, as a framework or approach, has a history of adaptation; starting with Rip and Kemp (1998) and Geels' (2002) work, and more recent interpretations such as Sangawongse et al. (2012) and Geels (2014), among others. The MLP can be thought of as wireframe for nested hierarchies where different labels can be affixed to it and additional components can be added to provide explanations. The evolutionary approach to studying transitions through the lens of the MLP, which favours niche pressure on the regime as a vehicle for transitions, provides a way to investigate a niche innovation within a system. The MLP is used to investigate the temporal and a structural scale of transitions; however, as Raven et al. (2012) have demonstrated, the spatial scale is not explicitly conceptualised. This research proposes viewing cities within their own place-bound MLP by adapting the framework to investigate urban transitions and modifying the three levels to reflect an urban system (city, urban planning, and a niche land use). An adapted MLP, whereby the regime reflects urban planning, focuses not only on spatial elements but also reveals the exercise of structuring activities and innovations within cities. An adapted MLP also helps to highlight the reliance on governance structures, approaches to urban planning, and key stakeholders to influence the transition process, which responds to Geels' (2014) call to study regime dynamics and not just niche innovations.

12.1.2 What Is Urban Planning?

Urban planning is inherently complex – there is more than one objective (economic growth, fair distribution of income, social cohesion and stability, reduction of psychological stress, a beautiful environment, etc.) – and the processes are multidimensional (Hall and Tewdwr-Jones 2011). As Rittel and Webber (1973, p. 136) contended, 'planning problems are inherently wicked'. Urban planning is the practice by which plans, programs, and designs are developed to intervene in the built environment in a quasi-evolutionary approach. Rather than have cities evolve blindly, planners set a course to move towards a goal, such as a more sustainable, resilient, compact, integrated, equal, or just future.

Urban planning is a spatial exercise. It can be seen as a method to assume control over space and human activity or the practice 'to connect forms of knowledge with forms of action in the public domain' (Friedmann 1993). Land-use planning (also known as physical planning or spatial planning) 'refers to planning with a spatial, or geographical, component, in which the general objective is to provide for a spatial structure of activities (or and uses) which in some way is better than the pattern existing without planning' (Hall and Tewdwr-Jones 2011, p. 3). At a more practical level, urban planning is responsible for granting approval of developments (Denyer-Green 2012). There are a number of mechanisms employed by local governments and urban planning departments to control land use. These include zoning, development controls, design guidelines, and building codes, amongst others (see also Chap. 4).

In urban planning, it has long been argued there are theories *of* planning (why it exists and what it does) and theories *in* planning (how to go about it) (Allemendinger 2002; Faludi 2013). The history of urban planning highlights reasons why planning exists (such as the need for effective government administration over land use) and what planning does (to provide structure and control over space and human activity). Different theories or approaches are chosen by planners for a myriad of reasons (conflicting pressures including personal/professional, place of work, location, professional body, and education, amongst others), and sometimes different approaches are required for different circumstances (Allemendinger 2002). Two theories in planning that are particularly relevant to this chapter are physical development and communicative planning. Physical development planning was the domain of architects and engineers, as well as utopian plans for cities. Modernity and the industrial revolution influenced planning and created interest in building and land regulations, such as property rights, zoning, and master plans (Healey 2012). This evolved to wanting to understand urban form and human settlements and how to manipulate it (Moudon 1997). Communicative planning is a more open and participatory planning process. It is a method for expanding institutional capacity and helping local communities to compete with global political-economic forces and direct foreign investment (Healey 1998).

12.2 Analytical Framework: Social-Ecological Transitions for Urban Planning

This research began with an interest in small-scale interventions within the built environment and was drawn to the MLP as a way to structure the investigation. However, the MLP needed to be adapted to better suit the aims of this research. First, the MLP, which is primarily used to study social-technical transitions (STT), was adapted to investigate transitions within the context of cities as social-ecological

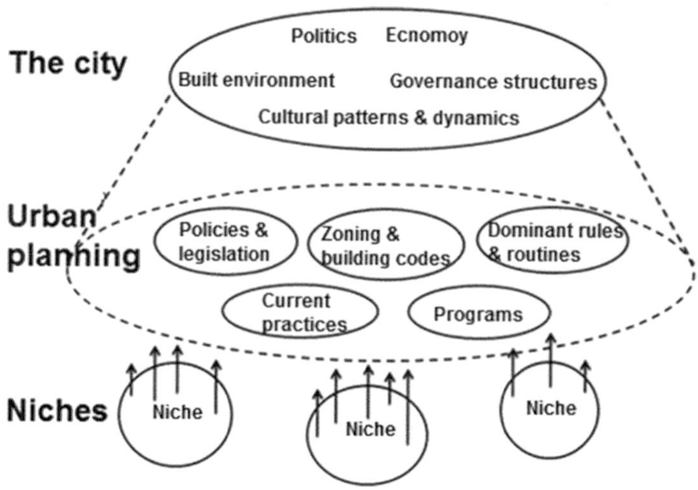

Fig. 12.1 Social-ecological transitions for urban planning (Source: author, adapted from Geels 2002)

systems (SES). SES, rather than STT, is used as the theoretical perspective because cities are complex systems made of both the human (which includes technology) and biophysical (Gallopín 2006; Walker and Salt 2006; Davoudi et al. 2013). Gallopín et al. (2001) argued that SES are the natural analytical unit for sustainable development. Smith and Stirling (2010) described SES as being 'place bound' and are situated within a spatial context such as a watershed, rangeland, forest, or region. Then the three levels of the MLP were adapted to reflect an urban system: the landscape becomes the city, the regime urban planning, and niches vary depending on the city and context. Figure 12.1 illustrates the application of a MLP for urban transitions. This approach will be referred to as Social-Ecological Transitions for Urban Planning (SETUP). SETUP provides a structure to investigate the trajectory and acceleration of particular niche innovations within cities[1,2]. By adapting the MLP to have the regime reflect urban planning, it focuses not only on the spatial element but also on the exercise of structuring activities and innovations within cities. It also helps to highlight the reliance on governance structures, approaches to urban planning, and key stakeholders to influence the transition process, which responds to Geels' (2014) call to study regime dynamics and not just niche innovations.

[1]The author recognises that by using cities as a bounded system and adjusting the scale of the landscape to describe a city, other external factors that impact the city are potentially missed, such as dimensions located within larger regional contexts such as state/province, national, as well as global contexts.

[2]The author acknowledges that a limitation to using the MLP to track the trajectory of niche innovations is the oversimplification of the levels.

Within the MLP, the landscape represents the highest level within the hierarchy (Geels 2002). This means that transformations initiated from the landscape should be viewed as top-down and that changes at the landscape level can create pressure or influence changes on both the regime and niches (Kemp et al. 2001). The city, as the landscape, is relatively static and cannot be easily changed by actors in the short term. The landscape embodies the physical, technical, and material setting that supports the city as a system. The landscape level of investigation includes the overall setting and relevant history of each place. Understanding the geographic location, politics, economics, governance, and culture of the city offers a setting for the nesting of the regime and niche levels of the city.

The regime level represents a system's rules and regulations (Geels 2002). It is comprised of the structures that support the current practices and routines, including the rules that provide stability and reinforcement to a system (Schot and Geels 2008). It exists in a shorter time frame than the landscape level and is more malleable. In the cycle of adaptive change, a disturbance to the system, such as a creative destruction, signifies a regime change. This is followed by a phase of renewal and reorganisation. Change in regime can come from the landscape level or be initiated from the niche level (Markard and Truffer 2008). The regime acts as the vehicle for change to move from one level to the next, be it top-down or bottom-up. Within the context of this research, the city as a social-ecological system, and the regime is represented through urban planning (and government). Therefore, urban planning, as the regime, is concerned with the structure, current practices, dominant rules (including zoning and building codes), and routines of the city.

Niches can influence transformation and represent experimentation and innovation (Geels 2002; Smith 2007). Niches, as small-scale interventions and radical innovations, build up internal momentum, which may lead to bottom-up change (Markard and Truffer 2008; Schot and Geels 2008). Moving upwards in the nested hierarchy of the MLP, niches may change the regime, and a new regime changes the landscape over the long term. The absence of structure and coordination at the niche level, compared to the regime and landscape, allows for new interactions to take place that may support innovation.

12.3 Methodology

This research compares the development and adoption of live/work as a niche in two different cities: Vancouver and Melbourne. An embedded multiple-case-study approach with a theoretical replication design was used for this research; each individual case has the same central analytical framework and the same questions applied to it. This approach was chosen to explore contrasting results (Yin 2009) in different cities. The aim of the case studies is to examine live/work in the two cities and to identify the process of adoption or transition of a niche innovation to a mainstream or regime policy. SETUP is used to provide a structure to investigate the trajectory and acceleration of niche innovations within different cities. This structure highlights the reliance on governance structures, approaches to urban planning, and stakeholders to influence the transition process.

Documentation was the primary source of data for this research and was supplemented by interviews, direct observation, and physical artefacts. Documentation in the form of policy research was used as the primary method (including each city's strategic planning documents, the government and urban planning department websites, relevant sustainability policies, and live/work policies and legislation) and was accompanied by studies of the cities, government and administrative material, media sources, and project information. Semi-structured interviews were conducted with urban planners, architects, developers, live/work occupiers, and an urban planning academic. Direct, unstructured observation in each city was conducted, and live/work units and developments served as physical artefacts.

12.3.1 Live/Work

The decline of manufacturing, a renewed interest in inner cities, and the adaptive reuse of vacant or underused buildings and lots paved the way for new environments and lifestyles (Roberts 2000; Bullen and Love 2009). In the 1970s, 'live/work' was used to describe the emerging trend of artists' loft spaces in New York (Dolan 2012; Hollis 2015). This trend emerged in San Francisco a decade later, and by the 1990s, most cities in North America and Western Europe had converted loft districts where people were potentially living and working (Christaanse 2012; Dolan 2012). In the absence of any other name, live/work began to be used as a generic term for buildings that combined dwelling and workplace (Hollis 2015). In this chapter, live/work is both a land use and a building type and is defined as a combined dwelling and workplace in a single unit or property. This definition recognises the unit or property is mixed use and supports the dynamic and adaptable nature of the functions over space and time.

Urban planning legislation that allows for combined dwelling and workplace employs terms such as 'mixed use', 'home occupation', and 'live/work' or 'work/live' to categorise or legalise the combined arrangement (Davis 2012; Dolan 2012). Mixed use blends residential, commercial, cultural, institutional, and, where appropriate, industrial uses together (APA n.d.). Combined dwelling and workplace are inherently mixed use (Davis 2012; Dolan 2012); yet they are simply a type of mixed use. Home occupation incorporates the right to pursue small-scale work activities at home. These types of arrangements are mostly known as a 'home office' or 'working from home' and usually have restrictions on employees and whether commercial/client visits are permitted (Dolan 2012). Live/work signifies that the dwelling is the primary use of the building or unit but that working is permitted and work/live means the needs of the work component take priority over the living (City of Vancouver 1996). In urban planning legislation, live/work is often associated with residential and mixed-use zones and codes, whereas work/live is associated with commercial, industrial, or mixed-use zones and codes (Dolan 2012). For the purposes of this chapter, live/work will be used to signify both live/work and work/live.

12.3.2 Case Studies: Vancouver and Melbourne

Two cities were chosen to compare contrasting experiences of a niche innovation; in Vancouver, live/work has been integrated into the regime, whereas in Melbourne, it has remained outside the system. The two cities also share similar development contexts over the past two decades (Holden and Scerri 2013). Holden and Scerri (2013) compared locations in each city for their similarities in expressions and symbols of the creative class, for example, the urban renewal developments of Yaletown in Vancouver and Southbank in Melbourne and the neighbourhoods of Commercial Drive or South Main in Vancouver and Brunswick's Sydney Road in Melbourne. However, one of the major differences between the two cities is the variety of housing types; both cities have detached, semi-detached, and apartments. In Vancouver, there are additional options, such as secondary suites, laneway houses, and live/work units. In Melbourne, secondary suites and laneway houses (also referred to as granny flats) exist. The exact rules vary from council to council, but typically proof that the occupier is a dependent person is needed. Victoria has some of the toughest regulations in Australia (Dow 2016).

A secondary suite is an additional (secondary) dwelling unit in a house (City of Vancouver 2013a). A laneway house is a smaller, detached home located where the garage would normally be located on a single-detached lot (City of Vancouver 2013b). In Vancouver, live/work is defined under six categories: commercial live/work, commercial work/live, artist live/work, artist work/live, industrial live/work, and industrial work/live (City of Vancouver 1996). These housing strategies have been used alongside different planning strategies, such as the Living First Strategy in the early 1990s, which created more housing in the central city and included artist studios and live/work. Another strategy was the City's EcoDensity Charter of 2008, which aimed to increase density through invisible density (secondary suites), hidden density (laneway houses), and gentle density (semi-detached and mixed-use housing).

Live/work has been connected to artist populations in New York, San Francisco, Oakland, and Vancouver, amongst many other cities (Zukin 1982; Baum and Christiaanse 2012; Dolan 2012). Like Vancouver, Melbourne also has artist communities blurring the boundaries of combining their living and working spaces for decades, but live/work as a formalised land use does not exist. However, as a building typology, live/work can be found informally in different parts of Melbourne, meaning they are harder to locate, demonstrating that the spatial impacts of unsuccessful niches are harder to measure. In Melbourne, home occupations are permitted in current residential and mixed-use zones, but only as long as they do not disturb the surrounding neighbours. The main difference between live/work and home occupation is the disturbance factor on neighbours. For example, live/work units are permitted to have employees and walk-in trade in their units, but this is not permitted for home occupations (City of Vancouver 2006).

12.4 Vancouver

The City of Vancouver is located on the West Coast of Canada. Vancouver was built upon a grid system and is the most densely populated city in Canada. In 2011, the City of Vancouver had a population of 603,502, while Metro Vancouver has a population of 2.3 million (Statistics Canada 2011). Metro Vancouver is surrounded by water and mountains and is known for its aesthetic beauty, scenic views, and mild climate. The Economist Intelligence Unit and Mercer's quality of life rankings continually rank Vancouver as one of the world's most liveable cities. Vancouver is governed by City Council and is part of Metro Vancouver, a regional government political body and corporate entity comprised of 24 local authorities. The City is also a charter city, meaning that it is regulated under the Vancouver Charter (City of Vancouver 2012). Its charter city status means that it does not need to prepare an official community plan (strategic planning document), which all other municipalities in British Columbia are legislated to do.

The City has a strong belief in separating politics from planning; therefore, decisions regarding major rezonings and redevelopments are delegated to the Development Permit Board (a technocracy chaired by the director of planning) and not by council. Vancouver is known for its urban planning and urban form, especially its high-rise residential and mixed-use development, in the form of the podium-tower, with commercial space on the ground floor and residential above along the city's transit corridors or within larger developments that mix retail and commercial with live/work and parking and loading. Vancouver approaches planning from a neighbourhood or district scale versus whole-city scale, which is permitted due to its charter city status. This approach supports a more experimental attitude towards mixing of uses, and there is evidence of the promotion of more housing choices. In practice, the City pursued live/work developments for the minimisation of travel time, as well as supporting artists, artisans, and small businesses. These changes not only created a greater supply of live/work units but also helped to dramatically increase the number of people living and working within the inner city (Vancouver Planner 2015).

After artist protests in the mid-1980s, the City in collaboration with local artists and architects developed a series of relaxations on zoning and building codes to allow dwelling units in conjunction with artist studios (Ostry 2015). The original live/work projects were intended for artists but became very popular with technology-based fields of the new economy, especially new media. In 1995, the City responded by approving changes to permit non-artist live/work (City of Vancouver 1996) and established three different live/work types (artist, commercial,

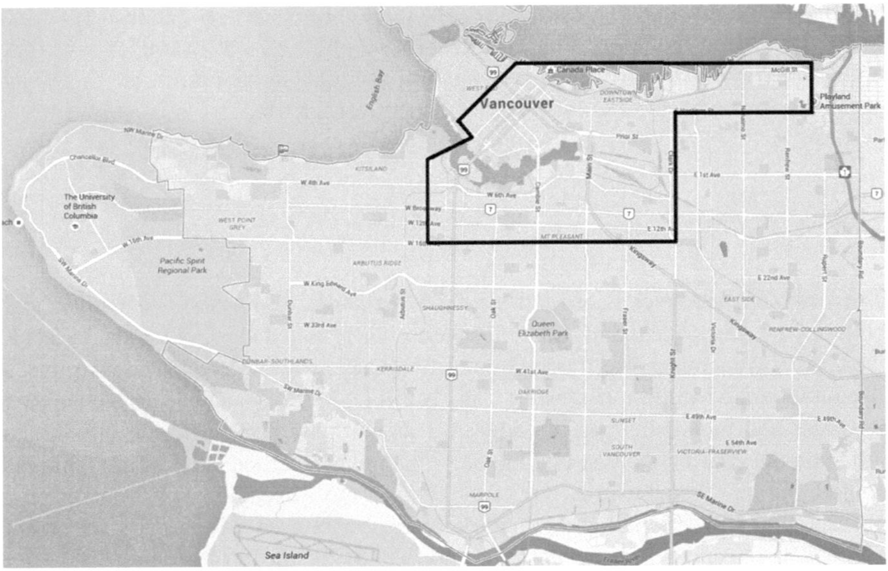

Fig. 12.2 Live/work in Vancouver (Googlemaps, adapted by author)

and industrial) to respond to the increased interest. The different types signified who the developments were meant for but also where they would be better situated within the city and specific zoning and development bylaws as well as building bylaws. In conjunction with the City's Living First Strategy, live/work developments were permitted in Brewery Creek, near Main St and Broadway, the IC3 district, and along Great Northern Way. These locations fall within three districts, Downtown, Downtown Eastside, and Mount Pleasant; later, more live/work units were permitted in Fairview. Live/work developments are permitted in other areas but are predominately found in the outlined area; see Fig. 12.2.

In 2006, Council adopted new Live/Work Use Guidelines that are still in operation today. However, at the end of the 2000s, Council put a hold on most new live/work developments. This was as a result of two main challenges (Vancouver Planner 2015). First, the move from rental to ownership meant that more people are living in their units and not living and working. Second, the way taxes are assessed in the province means that marketplace live/work units are assessed as residential (highest assessment value) but taxed as commercial (highest tax rate). In the case of live/work in Vancouver, rental has been much more resilient than its marketplace counterpart, because the owners tend to absorb the taxes and build it into the rent. A post-occupancy study conducted by the City also found that many of the occupiers were both living and working in rental live/work units (Vancouver Planner 2015).

The trajectory of live/work in Vancouver is illustrated in Fig. 12.3, and the timeline is provided in Table 12.1. Live/work in Vancouver accelerated and joined the regime but is currently in a state of stagnation due to barriers from the provincial

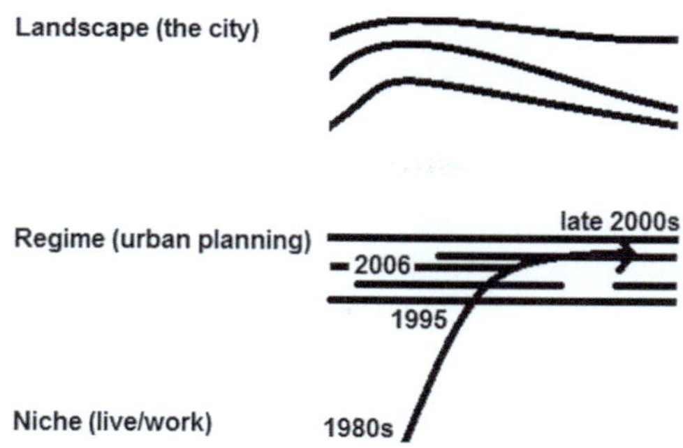

Fig. 12.3 The trajectory of live/work in Vancouver

Table 12.1 Time of live/work in Vancouver

Date	Policy	Action
Late 1980s	Development of live/work guidelines	Formalisation of artist live/work
1995	City Council approves non-artist live/work	Formalisation of non-artist live/work
2006	Live/work policies reviewed and updated	Update of artist and non-artist live/work
Late 2000s	City Council put a hold on most new live/work developments	Live/work developments come to a halt

governments, as well as issues regarding housing affordability. While live/work may not have contributed to a radical transformation of the city, it changed the regime level and encouraged people to think differently about development opportunities.

12.5 Melbourne

Melbourne is located in south-eastern Australia. In 2012, the population of Metropolitan Melbourne was 4.3 million (Victorian State Government 2014), and as of 2014, the City of Melbourne's (LGA) population was 122,207 (City of Melbourne 2016). Like Vancouver, it also continually ranks as one of the most liveable cities. Melbourne is known for its culture, sport, nightlife, and laneways. The region's economy is supported by the Port of Melbourne, the largest in the country, its manufacturing industries, professional services, and health, education, and social services (Victorian State Government 2014). Metropolitan Melbourne is a geographic designation; unlike Metro Vancouver, Metropolitan Melbourne does not have a regional level government authority (Gleeson et al. 2010). Rather than have one primary or large city within a metropolitan or regional authority, there are 31 Local Government Areas (LGA) in Metropolitan Melbourne, and each LGA is governed by a mayor and council (DTPLI 2015). However, the City of Melbourne is designated as the State of Victoria's capital city and is where a large proportion of the region's jobs are located (Spiller 2011).

Planning is administered by the Victorian State Government through the Department of Environment, Land, Water and Planning (DELWP) under the guidance of the Minister of Planning. The State Government provides statutory and strategic guidance on planning, while the LGAs are responsible for issuing planning permits and administering the planning scheme. The Planning Minister is the responsible authority for specific areas, and a number of statutory authorities report to the Minister, including the government's development and building regulation agencies (DTPLI 2014). This approach to urban planning is highly politicised. For example, 'Metropolitan Melbourne has endured five supposedly twenty to forty year time frame metropolitan strategies in as many decades (1971, 1980, 1995, 2002, 2014); none, thus far, has outlasted a change in state government (March 2012)' (Whitzman and Ryan 2014, p. 4). While the instability of politics at the state level is one barrier to effective planning, another barrier is that LGAs are not recognised in Australia's constitution, meaning their governance is weak (Gleeson et al. 2010). This affects the ability of local level governments to influence and lead urban planning decisions.

Unlike Vancouver, there is no live/work or work/live legislation in the state of Victoria. Under the current residential and mixed-use zones, home occupations are permitted in Metropolitan Melbourne, as long as they do not disturb the surrounding neighbours. Home occupation is defined as an occupation carried out in a dwelling or on the land around the dwelling by a resident of the dwelling. The Planning Scheme rules are the same for running a business out of a rental property or in an owner-occupied property, although individuals need to refer to their lease agreement or owner corporation's rules (City of Melbourne n.d.). Within the Victorian Planning and Environment Act 1987 (version incorporating amendments of 20 May 2013), residences for an occupier or caretaker of the shop, office, warehouse, or factory is permitted, but these spaces fall outside of the definition of dwelling. Using Dolan's (2012) live/work classifications, the occupier or caretaker residence would be referred to as live-near proximity type, which is essentially live/work with some separation between the living and working spaces.

The website creativespace.net.au lists a few live/work spaces available for rent. These spaces fall along the spectrum of lower rent artist spaces in warehouses, to polished tech-focused live/work spaces with commercial leases. Space Tank Studio and the Wardlow Artist Residence are both places that have hosted artists for 3-month artist in residence (artist live/work) in the past. 'Typically creative industries rely on councils with a "don't ask, don't tell" policy… If complaints are made by neighbours, councils investigate' (Professor Carolyn Whiztman, quoted in Edgar 2015, no pagination). However, the example of Sibling's Sleeping Pods, where two members of the architectural practice lived in the warehouse they were redesigning were eventually evicted by the council from their living quarters, provides another view of the current climate for live/work in Melbourne (Edgar 2015). In 2015, there were a series of discussions in newspapers, blogs, and public forums regarding spaces for artists in Melbourne and people living, and living and working in warehouses. However, it is yet to be seen whether the interest and momentum from the niche level will interact with, and possibly join, the regime.

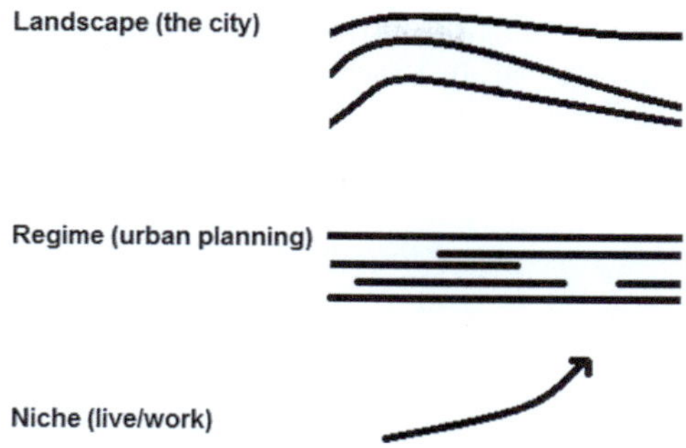

Fig. 12.4 The trajectory of live/work in Melbourne

Table 12.2 Case-study conditions

Conditions	Vancouver	Melbourne
Governance structures	Municipal, multilevel, charter city	Municipal (LGA), state government
Urban planning approaches	Physical development, communicative	Physical development
Stakeholders (initiators)	Artists, architects, developers, City Council	Artists, entrepreneurs
Stakeholders (Inhibitors)	City Council	Local Councils, state government

The trajectory of live/work in Melbourne is illustrated in Fig. 12.4. Live/work has not been formalised as a land use or dwelling type in Melbourne. People have and continue to occupy buildings or units as combined dwelling and workplace, but it is not legally recognised by the government.

Note: No map or timeline of Melbourne is provided because home occupation and caretaker residences are permitted throughout the metropolitan region, and true live/work is done informally, under the radar.

12.6 Discussion: Success Factors and Challenges

The trajectories of live/work in the different cities demonstrate how a particular niche accelerates and which policies, governance contexts, or stakeholders influence the process. The conditions of live/work as a niche intervention in Vancouver and Melbourne are summarised in Tables 12.1 and 12.2.

12.6.1 Governance Structures

Traditional forms of governance have been viewed as 'a bureaucratic state trying to impose its plan on society' (Bevir 2009, p. 22). Urban planning decisions are most often made within government institutions and in highly politicised environments.

The particular position of urban planning within the local government hierarchy is not universal. In Vancouver, it is the municipal government, and in some cases the metropolitan authority, which is responsible for urban affairs, as well as statutory issues. Under the Vancouver Charter, the provincial government granted the city some special powers to govern the city and provide services. In Melbourne, the state government through the DELWP is responsible for urban affairs and planning regarding long-term strategic planning and infrastructure. Implementation- and monitoring-based planning activities take place in the LGA planning departments. The case studies suggest that having a centralised urban planning approach within one level of government allowed Vancouver to be more collaborative and open to change whereas having two levels of government and multiple departments in Melbourne makes it more difficult. A reframing of planning to include a more reflexive (Davoudi and Strange 2009; Voβ et al. 2006) and collaborative (Healey 2012) approach could support a smoother transition to a more sustainable future.

12.6.2 Urban Planning Approaches

Particular urban planning approaches are selected for diverse reasons (Allmendinger 2002). Cities may operate under the umbrella of a particular approach, while sometimes different programs or parts of the city call for another perspective. In addition, individual planners bring their own worldview to their work, even if they are working within a particular approach presented in their city's vision or strategic planning document. Yet, in the case of the study cities, when it came to planning regarding live/work or other types of alternative developments, certain approaches in each city were made apparent. Vancouver's approach to urban planning represents a mix. Vancouver follows a physical development approach, whereby 'expert' planners make decisions on urban planning matters. The City also subscribes to a situation-specific approach, as seen with their neighbourhood plans (rather than a whole-city strategic plan). Vancouver also has a long history of community consultation, which was seen in the development of the initial live/work guidelines. Melbourne also follows a physical development approach. However, there are concerns over fragmentation and the transparency of the decision-making processes (Legacy et al. 2014), as well as the power of certain stakeholders (such as complaining neighbours). When an urban planning department and city council are more receptive and open to working with and listening to community members, the dialogue allows for new ideas to be presented and eventually permitted. This approach in Vancouver appears to be stronger when urban planning and politics have some separation.

12.6.3 Stakeholders

Support for live/work in each city first came from grassroots/bottom-up movements involving artists and their allies. For live/work to move from the niche space into the regime, it needs to achieve recognition and acceptance by a particular person or department within the government; without the intervention of the government, live/work would continue to be illegal. Stakeholders from inside and outside the regime helped live/work takes off in Vancouver and included artists, architects, developers, urban planners, and the City Council. However, it was the City Council that had the decision-making power to formalise live/work as a land use. Unfortunately, as a result of concerns over affordability due to provincial tax regulations and more individual ownership than rental developments, the City Council has approved far less live/work developments than in the past. In Melbourne, decisions to formalise a new zoning or residential type are under the authority of the state government.

12.7 Emerging Theoretical Space: Urban Planning and Sustainability Transitions

To date, the study of sustainability transitions has been dominated by sociotechnical transition scholars and economists (Bernstein and Hoffmann 2015). Bernstein and Hoffmann (2015) believe this presents an opportunity for those in other fields to contribute to theoretical innovations in the study of transitions. There has been a movement to go beyond seeking innovative technological solutions to pursuing methods that include spatial planning and urban policy (Bruyninckx 2015). Since a large proportion of greenhouse gas emissions in Australia and Canada come from buildings and transportation (EPA Victoria 2011; EC Canada 2014), it is important to understand urban form factors such as access to services and jobs, provision of public transport, and land-use mixes that impact greenhouse gas emissions in cities (Han et al. 2015). While other disciplines examine what happens within cities and why, urban planning uses this information to intervene and shape cities.

Urban planning and transitions research share a number of similarities. 'Transitions involve mutually coherent changes in practices and structures, and because of their multilayeredness and inevitable entrenchment in society and culture at large they are complex and comprehensive phenomena' (Grin et al. 2010, p. 3). Transition research is concerned with the capacity to accelerate and guide social innovation processes towards a more sustainable future (Loorbach and Rotmans 2010), and urban planning is about intervening in human settlements to achieve certain environmental, social, or political objectives (UNHSP 2009). Transitions and urban planning both share the desire to shape systems towards a specific normative goal. In the case of transitions, the orientation is sustainable development, and with urban planning, it is towards a better city.

The SETUP framework emphasises the role of urban planning and urban governance for sustainable transitions to occur at the city or regional scale, especially when the niches in question are not new technologies but changes in infrastructure or the built environment or policies to support sustainable practices or changes. SETUP supports the views of researchers advocating for the inclusion of the spatial within sustainability transitions research because urban planning is a spatial exercise, where the goal is to provide spatial structure to activities within cities.

12.8 Conclusion

The purpose of this research was to analyse the policy adoption or transition pathways of live/work in Vancouver and Melbourne. The two case-study cities are both located within complex governance structures with multiple scales of government authorities and multiple layers of decision-making, regulations, and policies. I identified that rigid and top-down governance structures are less flexible and open to change, political approaches to urban planning are less responsive and adaptive, and strong political actors have the ability to either initiate or inhibit change. Meanwhile, collaboration and experimentation with technical skill support the acceleration of niche innovation. The experience of live/work as a niche in each case study provides insight into the processes of acceleration across the different geographical, political, and cultural contexts. It also contributes to the process of identifying and highlighting challenges and success factors in the adoption or failure (possibly only temporary) of these policy innovations. These findings can also contribute to the development and adoption of live/work policies in other cities or to the development of other niche interventions land-use policies.

References

Allemendinger P (2002) Planning theory. Palgrave, New York

Alexander C, Ishikawa S, Silverstein M, Jacobson M, Fiksdhal-King I, Angle S (1977) A pattern Language: towns, buildings, construction. Oxford University Press, Oxford

American Planning Association (n.d.) Zoning for Mixed Uses. PAS QuickNotes, vol 6. American Planning Association

Baum M, Christaanse K (2012) City as Loft, adaptive reuse as a resource for sustainable urban development. GTA Verlag, Zürich

Bernstein S, Hoffman M (2015) The politics of decarbonization: a framework and method, Working paper. http://papers.ssrn.com/sol3/papers.cfm?abstract_id=2619322. Accessed 9 June 2016

Bevir M (2009) Key concepts in governance. Sage, London

Bruyninckx H (2015) Sustainability transitions and EU policy, Sustainability transitions and wider transformation change: historical roots and future pathways. In: J Schot (ed) Sustainability Transitions Research Network, July 25

Bullen PA, Love PED (2009) Residential regeneration and adaptive reuse: learning from the experiences of Los Angeles. Struct Surv 27(5):351–360
Christiaanse K (2012) Traces of the city as Loft. In: Baum M, Christaanse K (eds) City as Loft, adaptive reuse as a resource for sustainable urban development. GTA Verlag, Zürich, pp 14–24
City of Melbourne (2016) Acts about Melbourne. City of Melbourne. http://www.melbourne.vic.gov.au/about-melbourne/melbourne-profile/pages/facts-about-melbourne.aspx. Accessed 25 Apr 2016
City of Melbourne (n.d.) Do I need a planning permit? City of Melbourne. https://www.melbourne.vic.gov.au/BuildingandPlanning/Planning/FAQ/Pages/DoIneedaplanningpermit.aspx. Accessed 2 June 2015
City of Vancouver (1996) Live/work and work/live: Vancouver overview including strategic directions, City Of Vancouver
City of Vancouver (2006) Live/work use guidelines. City of Vancouver, Vancouver
City of Vancouver (2012) The Vancouver charter. City of Vancouver. http://vancouverca/your-government/the-vancouver-charteraspx. Accessed 12 May 2015
City of Vancouver (2013a) Secondary suite program, City of Vancouver
City of Vancouver (2013b) Laneway housing how to guide, City of Vancouver
Coenen L, Truffer B (2012) Places and spaces of sustainability transitions: geographical contributions to an emerging research and policy field. Eur Plan Stud 20(3):367–374
Davis H (2012) Living over the store, architectural and local urban life. Routledge, New York
Davoudi S, Strange I (2009) Conceptions of space and place in strategic spatial planning. Routledge, London
Davoudi S, Brooks E, Mehmood A (2013) Evolutionary resilience and strategies for climate adaptation. Plan Pract Res 28(3):307–322
Denyer-Green B (2012) Development and planning law. Taylor and Francis, Hoboken
Department of Transport, Planning and Local Infrastructure (2014) About planning in DTPLI. Victoria State Government. http://www.dtpli.vic.gov.au/planning/about-planning/about-planning-in-dtpli. Accessed 2 June 2015
Department of Transport, Planning and Local Infrastructure (2015) Find your council. Victoria State Government. http://www.dtpli.vic.gov.au/local-government/find-your-local-council#councils. Accessed 2 June 2015
Dolan T (2012) Live-work planning and design: zero-commute housing. Wiley, Hoboken
Dow A (2016) Granny flats: a cheap housing solution? The Age, 01/21/2016
Edgar R (2015) What housing crisis? A new breed of nomads creates a cultural shift, The Sydney Morning Heard, June 5
Environment and Climate Change Canada (2014) Greenhouse gas emissions by economic sector. environment and climate change Canada, Ottawa, ON
Environment and Protection Agency Victoria (2011) Households and GHG emissions, environment and protection agency Victoria, Melbourne, VIC
Faludi A (2013) A reader in planning theory. Elsevier Science, Oxford
Friedmann J (1993) Toward a non-Euclidian mode of planning. J Am Plan Assoc 59(4):482–485
Friedman A (2012) Fundamentals of sustainable buildings. Island Press, Washington, DC
Gallopín GC (2006) Linkages between vulnerability, resilience, and adaptive capacity. Glob Environ Chang 16:293
Gallopín GC, Funtowicz S, O'Conner M, Ravetz J (2001) Science for the twenty-first century: from social contract to the scientific core. Int Soc Sci J 53(168):219–229
Geels FW (2002) Technological transitions as evolutionary reconfiguration processes: a multi-level perspective and a case-study. Res Policy 31:1257–1274
Geels FW (2014) Regime resistance against low-carbon transitions: introducing politics and power into the multi-level perspective. Energy Soc 0:1–20
Gibbs D, O'Neill K (2014) Rethinking sociotechnical transitions and green entrepreneurship: the potential for transformative change in the green building sector. Environ Plann A 46:1088–1107

Gleeson B, Dodson J, Spiller M (2010) Metropolitan governance for the Australian city: the case for reform issues paper 12, Urban Research Program, Griffith University
Grin J, Rotmans R, Schot J (2010) Transitions to sustainable development: new directions in the study of long term transformative change. Routledge, New York
Hall P, Tewdwr-Jones M (2011) Urban and regional planning, 5th edn. Routledge, New York
Han SS, Green R, Wang MY (2015) Towards low carbon cities in China: urban form and greenhouse gas emissions. Routledge, New York
Hansen T, Coenen L (2015) The geography of sustainability transitions: review, synthesis and reflections on an emergent research field. Environ Innov Soc Trans 17:92–109
Healey P (1998) Building institutional capacity through collaborative approaches to urban planning. Environ Plann A 30(9):1531–1546
Healey P (2012) Traditions of planning thought. In: Fainstein SS, Campbell S (eds) Readings in planning theory. Wiley-Blackwell, Hoboken
Hodson M, Marvin S (2010) Can cities shape socio-technical transitions and how would we know if they were? Res Policy 39:477–485
Holden M, Scerri A (2013) More than this: liveable Melbourne meets liveable Vancouver. Cities 31:444–453
Hollis F (2015) Beyond live/work: the architecture of home-based work. Routledge, London
Kemp R, Rip A, Schot J (2001) Constructing transition paths through the management of niches. In: Garud R, Karnøe P (eds) Path dependence and creation. Lawrence Erlbaum Associates, London, pp 269–299
Legacy C, Gleeson B, Dodson J (2014) Implementation: getting our act together, In: Whitzman C, Gleeson B, Sheko A (eds) Melbourne: what next?, Research Monograph No. 1, Melbourne Sustainable Society Institute, The University of Melbourne, pp 116–131
Loorbach D, Rotmans J (2010) The practice of transition management: examples and lessons from four distinct cases. Futures 42:237–246
Markard J, Truffer B (2008) Technological innovation systems and the multi-level perspective: towards an integrated framework. Res Policy 37(4):596–615
Moudon AV (1997) Urban morphology as an emerging interdisciplinary field. Urban Morphol 1(1):3–10
Murphy JT (2015) Human geography and socio-technical transition studies: promising intersections. Environ Innov Soc Trans 17:73–91
Newman P, Jennings I (2008) Cities and sustainable ecosystems: principals and practices. Island Press, Washington, DC
Ostry M (2015, January 9) Personel communication
Pickett STA, Cadenasso ML, Grove JM (2003) Resilient cities: meaning, models, and metaphor for integrating the ecological, socio-economic, and planning realms. Landsc Urban Plan 69:369–384
Raven R, Schot J, Berkhout B (2012) Space and scale in socio-technical transitions. Environ Innov Soc Trans 4:63–78
Rip A, Kemp R (1998) Technological change. In: Rayner S, Malone EL (eds) Human choice and climate change. Vol. II, Resources and technology. Battelle Press, Columbus, pp 327–399
Rittel HW, Webber MM (1973) Dilemmas in a general theory of planning. Policy Stud 4(2):155–169
Roberts P (2000) The evolution, definition and purpose of urban regeneration. In: Roberts P, Sykes H (eds) Urban regeneration, a handbook. Sage, Londan
Sangawongse S, Sengers F, Raven RPJM (2012) The multi-level perspective and the scope for sustainable land use planning in Chiang Mai City. Environ Nat Resour J 10(2):21–30
Schot J, Geels FW (2008) Strategic niche management and sustainable innovation journeys: theory, findings, research agenda, and policy. Tech Anal Strat Manag 20(5):537–554
Smith A (2007) Translating sustainabilities between green niches and socio-technical regimes. Tech Anal Strat Manag 19(4):427–450
Smith A, Sterling A (2010) The politics of social-ecological resilience and sustainable sociotechnical transitions. Ecol Soc 15(1):11

Spiller M (2011) Place making, inclusion and governance in the suburban city – a case study of Melbourne, Australia. In: Chisholm S (ed) Investing in better places: international perspectives. The Smith Institute, London, pp 80–95
Statistics Canada (2011) Population
Truffer B, Coenen L (2012) Environmental innovation and sustainability transitions in regional studies. Reg Stud 46(1):1–21
Truffer B, Murphy JT, Raven R (2015) The geography of sustainability transitions: contours of an emerging theme. Environ Innov Soc Trans 17:63–72
Vancouver Planner (2015) Personal communication. 8 January
UNHSP (2009) Planning sustainable cities, global report of human settlements 2009, United National Human Settlement Programme, Earthscan, London
Victorian State Government (2014) Plan Melbourne: metropolitan planning strategy. Victorian State Government, Melbourne
Voβ JP, Kemp R, Bauchnecht D (2006) Reflexive governance: a view on an emerging path. Teokessa Voss, JP
Walker B, Salt D (2006) Resilience thinking: sustaining ecosystems and people in a changing world. Island Press, Washington, DC
Whitzman C, Ryan C (2014) A Vision for Metropolitan Melbourne. In: Whitzman C, Gleeson B, Sheko A (eds) Melbourne: What Next? Research Monograph No. 1, Melbourne Sustainable Society Institute, The University of Melbourne, pp 4–20
Yin R (2009) Case study research: design and methods, 4th edn. SAGE Inc, London
Zukin S (1982) Loft living. The John Hopkins Press Ltd, London

Chapter 13
The Socioeconomic Equity Dimensions of a Transition in Suburban Motor Vehicle Fuel and Technology

Jago Dodson, Terry Li, and Neil Sipe

Abstract This chapter investigates the social and spatial equity implications of a transition to high fuel efficiency fossil-fuel vehicles or to non-fossil-fuelled vehicles for urban travel in Australian cities. The chapter draws on empirical work undertaken by the authors that reveals that the advantages of high fuel-efficient vehicles will largely be disproportionately captured by wealthier households. Given the spatial structure of Australian cities, these households also typically reside in areas well served by public transport and where cycling and walking are relatively more prevalent. The consequences of this connection between technology, socioeconomic patterns and urban structure are that a transition to high fuel-efficient vehicles will likely have adverse socioeconomic consequences for highly car-dependent low income households in the outer suburbs of Australian cities. Policy that can better manage the transition to a lower carbon urban transport system through more systemic reform than market-led vehicle fuel efficiency improvements will be needed if we are to avoid regressive socioeconomic outcomes. The chapter will place this discussion within the context of the wider transitions literature.

Keywords Mobility • Transport • Socioeconomic equity • Fuel efficiency • Suburbs

J. Dodson (✉) • T. Li
Centre for Urban Research, RMIT University, Melbourne, VIC, Australia
e-mail: jago.dodson@rmit.edu.au; tiebei.li@rmit.edu.au

N. Sipe
School of Earth and Environmental Sciences, Faculty of Science, University of Queensland, Brisbane, QLD, Australia
e-mail: n.sipe@uq.edu.au

T. Moore et al. (eds.), *Urban Sustainability Transitions*, Theory and Practice of Urban Sustainability Transitions, DOI 10.1007/978-981-10-4792-3_13

13.1 Introduction

Urban transport systems are among the most highly anticipated sectors to experience major technological and organisational disruption in the coming decades, with considerable potential for transition to new technological and institutional arrangements. This potential transition is especially prominent in the extensive suburban areas found in many of the cities of North America and Australasia in which high levels of reliance on private motor vehicles for transport are associated with a dispersed pattern of land use and a household tenure structure that favours private home ownership.

Although there are variations between the North American and Australasian cases, in general dispersed car-dependent suburbia also differentiates households according to wealth and income. In Australia less affluent urban households tend to be located in dispersed outer suburban areas. Many have been drawn there by relatively affordable detached dwellings which remain a favoured form for owner-occupiers within Australia's national culture of housing. That spatial tendency reflects the social history of post-WWII Australia in which the provision of good quality housing to the majority of households was achieved via suburban expansion of owner occupied detached housing, overwhelmingly served by the automobile. Although Australian cities have relatively dense public transport networks in central areas, these have not been systematically extended to post-WWII suburban areas and in general fail to offer a viable alternative to automobile travel for suburban residents (Newman and Kenworthy 1999; Mees 2010). For those in less affluent income tiers, car dependence is a sine qua non of affordable suburban home ownership.

Dispersed, detached suburbia, whether in Australasia or North America, now faces some major endogenous questions. Consumption of agriculturally valuable and environmentally sensitive peri-urban land, increasing spatial mismatch with employment and public services and rising costs of infrastructure provision have brought increasing criticism to the continued rollout of new suburban development at the fringe of cities (as discussed by Newton in Chap. 9). A further critique has assessed the policy failures of inadequate public transport provision to new suburban development in terms of the vulnerability that the resulting car dependence creates for suburban residents.

The concerns about Australian suburban form, structure and function sit within a wider exogenous reconsideration of suburbanisation generally over the past decade in which perceptions of car-dependent suburbia have shifted from an idealised form of human settlement to a morbid form (Gonzalez 2009; Mees 2010). This reappraisal rests upon perceived reliance of car dependence suburbia on climate harming and depleting liquid fossil fuels, as well as for its perceived effects on human health through the reduced rates of physical activity resulting from extensive car use (Hirschhorn 2005). The problem of climate harm and fossil fuel dependence have been systematically investigated, with an array of both academic analysis and popular commentary claiming that there is a need to shift away from conventional suburban configurations of automobile dependence (Newman and Kenworthy 1999; Mees 2010).

Strategic planning policy has however been slow to respond to suburban doubt by modifying the post-WWII model of development. While land-use strategies in Australia have included reference to more compact forms of urban development, this has also been accompanied by continued expansion of greenfield development and extension of road capacity and support for private automobile travel. Although there has been support within Australian metropolitan plans for increased development within existing urban areas, this has tended to be left to market processes to deliver such that new development tends to locate in higher density zones rather than in detached car-dependent suburbia. As Dodson (2012) has argued, most gains in urban density in Australian cities are happening in the wrong place to change travel behaviour or are occurring over time frames that are too slow to respond to the immediate risks of suburban car dependence. In contrast macro-level policy has focused on increasing motor vehicle fuel economy as a means to reduce suburban reliance on climate harming globally insecure fossil fuels. Indeed this approach has been adopted at the highest levels of global policy making with, for example, the White House Energy Policy (2011) proposes '... mak(ing) it easier and more affordable for consumers to buy more advanced and fuel-efficient vehicles...' (p. 4) as a principal means to reduce household (transport) energy costs. Australian national energy policy has tended to focus on market signals facilitating demand for increased motor vehicle efficiency rather than a deliberately targeted government strategy to reduce car dependence in cities.

The apparent reliance on more fuel-efficient vehicles as a means of transitioning from carbon intensive and supply-vulnerable fossil fuels raises questions about the viability of this transition and its effects. Various studies of car dependence in Australian suburbia have shown that it is the less affluent households that are more vulnerable to the adverse consequences of a fuel price shock or carbon-intensity-based pricing for motor vehicle fuels (Dodson and Sipe 2007, 2008). Yet a strategy premised on the inclusion of progressively more fuel-efficient technology into the Australian urban vehicle fleet would imply the purchasing of new vehicles, which are typically expensive relative to existing second-hand vehicles. This in turn raises questions about the social equity of an efficiency-based suburban transport energy transition. Simply stated, will vulnerable suburban households have access to the more fuel-efficient vehicles as these enter the market or will wealthier, but less car-dependent, households capture those vehicles? If state carbon regulation or global fuel market prices were to compel a more rapid shift, would the suburban fleet transition to a more efficient technological form? What is the role of social factors in enabling or hindering an accelerated motor vehicle fuel-technology transition?

This chapter investigates the challenges of a suburban transport fuel and technology transition founded on increased vehicle fuel economy as compelled by emerging global trends and forces. The chapter draws on work undertaken by the authors over the past decade that has sought to chart the contours of suburban oil vulnerability and to appraise the feasibility of transition strategies and strategies premised on maintaining automobility via increased vehicle fuel economy. The chapter argues that such strategies risk misunderstanding the dynamics of suburban land use and transport systems to the extent that they will fail to adequately prepare Australian suburbia for the necessary transition to a less climate risky, petroleum secure energy.

13.2 Regimes, Transitions and Peaks

The field of urban transport is widely perceived as having considerable potential for a future reconfiguration, possibly towards more sustainable travel modes. A major reason that such a transition in urban transport seems feasible is that this area of human activity has witnessed a series of marked transitions over the past two centuries. Until the early twentieth century, most human land-based travel, whether urban or not, relied on pedestrian or horse-drawn transport. The invention and expanding provision of railways for intra- and inter-urban mobility in the mid-nineteenth century permitted a major transition in the means, scale, frequency and distance of travel available to urban populations. The railway also had spatial consequences in reducing the time taken to traverse urban space and thus making practically feasible the expansion of cities beyond existing temporal bounds as set by the speeds and convenience of pedestrian or horse-drawn travel. Modern mass suburbia, though it had emerged in the horse-drawn phase of the industrial city (Broadbent 1987), was founded on railway travel (Davison 1979; Frost 2000). The arrival of the automobile during the early-twentieth century in turn provided the potential for a further urban mobility transition in North American and Australian cities especially following the expansion of freeway networks post-WWII. This transition had distinct spatial consequences. Rail transport tends to create concentrate high levels of accessibility into spatial nodes, such as stations, which are attractive to land uses. Automobile-based transport in contrast disperses accessibility and in turn makes marginal locations in centralised spatial land markets attractive to land uses. The spatial history of Australian and North American cities since WWII is largely one of the automobile-based suburban expansion (Jackson 1987; Davison and Yelland 2004).

This spatial factor is often overlooked in analyses of mobility and technology transitions, including how sociotechnical regimes can be formed (Coenen et al. 2012). The multi-level perspective, which is prominent in transitions debates (Cohen 2012) and uses the explicitly spatial metaphor of 'landscape', for example (Geels and Schot 2007, p. 401), attends closely to hierarchised institutional dynamics but dedicates little attention to spatial scales, either in terms of the spatio-institutional level at which transitions occur or the material spatial distribution of various sociotechnical formations. Take, for example, Geels' (2005) study which investigates the transition from horse-drawn transport to automobile-based transport over the period from 1860 to 1930. Although Geels (2005, fig. 1, p. 446) recognises that transport mode transitions were associated with the emergence of suburbs, his analysis dedicates relatively little attention to questions of land use and space generally.

The absence of a systematic spatial perspective in urban investigations of sociotechnical transitions seems peculiar given other 'regime-' oriented studies utilising a political economy perspective have attested to the very specific connections between the technological reconfigurations and the spatial reconfigurations embedded in the emergence of modern suburbia (Walker 1981; Harvey 1985; Florida and Jonas 1991). For example, Walker (1981) has described how suburbia in the USA comprises a 'spatial fix' to the contradictions of industrial capitalism in which the concentrated production and accumulation of capital in industrial cities was accompanied by immiseration of the working class in dilapidated and unsanitary housing and the wider pervasion of disease and pollution across the city. The spatial fix to this contradiction in capitalism was, for Walker, a resort to rail transport technology to compress urban distance and enable housing to be developed on ex-urban land while still connected to central production sites. In the post-WWII period, the term Fordism was used to describe the road-based version of this regime, which comprised a spatial relocation of industrial production to suburban sites linked via the automobile sociotechnical complex to housing systems organised around dispersed suburbia (Winter and Bryson 1998). This political economy approach to understanding suburbia has been applied in the Australian context by Berry (1984, 1999) to understand the development of Australian suburbia in which a similar use of automobile technology paired with the configuration of industrial production and stable redistributive social structures combined to entrench suburban home ownership as the dominant tenurial and socio-spatial form of Australian cities. And as Dodson and Sipe (2009) have noted in the context of the US sub-prime mortgage crisis, financial arrangements must be treated as a component of suburban configurations as these provide the basis for housing development premised on the spatio-technological accessibility of suburban land. Indeed Dodson (2014) has argued that suburbia should be viewed as an integrated and complex sociotechnical assemblage that combines land, housing technology, transport technology, energy supply networks and credit systems. Suburbia is configured through more than simple land-use relationships. Disruptions to elements of this complex in turn imply disruptions to the whole. For post-WWII suburbia, this includes wider disruptions to automobility and resistant efforts to sustain automobility in the face of disruption.

In addition to posing fundamental questions about the future of suburbia, the question of a sociotechnical transition in automobile fuel and technology also sits within the context of wider debates about the status of the automobile and its role in cities. From the late-2000s, a vigorous discussion emerged which queried whether increased fossil fuel prices or carbon emissions abatement costs would disrupt or reduce travel by automobile (Newman and Kenworthy 1999; Dodson and Sipe 2008; Gilbert and Perl 2008; Li et al. 2013). Evidence of such a reduction was limited though, until Puentes and Tomer (2008) described a plateau and decline in US car use during the 2000s. Metz (2010) subsequently observed a comparable 'saturation' in the use of cars for travel in the UK. Millard-Ball and Schipper (2011) and Kuhminhoff et al. (2013) subsequently described a plateau or decline in the growth in car travel among a number of developed countries, including Australia where car use has plateaued since 2005 and the USA (van Dender and Clever 2013, p. 6). Many researchers have subsequently come to view this change as an epochal transition in urban travel (Lipschutz 2012; Banister 2013; Dennis and Urry 2013; Goodwin 2013; Metz 2013; Rigal and Rudler 2014). A peak in car use poses elemental questions for both transport policy and for the future of suburbia because car use consistently increased within the major urban areas of Australia and North America during most of the past six decades. Most urban transport plans in Australia or North America expect car use to continue and plan infrastructure to accommodate this growth. A reduction in car use, within a potential wider restructuring of urban travel away from cars, radically disrupts such assumptions and introduces new uncertainty that could lead to excessive spending on car infrastructure if realised.

The peak car phenomenon has spurred two streams of investigation. The first assesses factors underpinning restructuring of car use (Urry 2012; Delbosc and Currie 2013; Grimal et al. 2013; Headicar 2013; Kuhnimhof et al. 2013). A second stream investigates the technical, economic, social and policy consequences for cities (Schipper et al. 2010; Cohen 2012). Not all of this work has been conceived from a theoretically oriented 'transitions' perspective, but the questions posed are closely aligned to the transitions stance. Little of this work has been spatially sensitive at the metropolitan scale though. There is a need for appraisal of the socio-spatial patterns of changing car use in cities, particularly in relation to the uptake of more fuel-efficient vehicle technology has been relatively modest. The vehicle fuel efficiency (VFE) of national motor vehicle fleets has been a major area of inquiry within debates over the future technological and fuel mix of future vehicle fleets. Most VFE research investigates the technical efficiency of particular vehicle technology and means of increasing performance (An et al. 2011; Daly and Gallachóir 2011; Schiraldi 2011; Sheinbaum-Pardo and Chávez-Baeza 2011; Hulsmannm 2012; Tyfield and Urry 2012; Brand et al. 2013). A further set of studies assesses the implications of aggregate national gains for total fuel consumption or transport carbon emissions (Gallagher and Muehlegger 2011; Graham-Rowe et al. 2011; Kagawa et al. 2011; Kloess and Müller 2011; Anable et al. 2012; Daly and Gallachóir 2012; Liddle 2012; Brand et al. 2013; Glerum et al. 2013; Liu and Santos 2013). Often such research includes policy discussion as to how to improve aggregate outcomes. This research attention is reflected in policy—President Obama's *New Energy Plan*, for example, seeks VFE gains in the US motor vehicle fleet as a means of reducing cost pressures on car-dependent US households.

Although policy research has assessed overall economic savings from improved VFE performance (Climate Change Authority 2013), distributional questions concerning VFE gains have been less considered. The household-level implications of increasing VFE or to the way that access to high efficiency vehicles is distributed in cities have received modest attention. Bhat and Guo (2007) and Guo and Bhat (2007) showed that household vehicle type is related to socioeconomic and spatial factors, at an individual level. Relatively little research however has investigated the direct relationship between the VFE of vehicle fleets and spatial socioeconomic patterns in large urban areas. This is an important gap in the transport literature because individual household experience of motor vehicle fuel efficiency gains is shaped by local sub-metropolitan conditions and wider distributional patterns in urban systems. Given there is evidence that a major policy response to increased fuel prices and carbon abatement imperatives among car-dependent jurisdictions such as the USA and Australia will be focused on transitioning to low emission motor vehicle fuel technology, there is an imperative to understand how this transition will occur and how it will be distributed. Questions as to the appropriateness of this policy direction, its feasibility and its distributional consequences demand a response. The remainder of this chapter responds to this problem.

13.3 Vehicle Fuel Efficiency in Australia

Carbon emissions from motor vehicles are directly related to fuel consumption. The most recent measurement was in 2011 at which point average private passenger vehicle fuel consumption in Australia was 11.1 litres per 100 kilometres (L/100km) (ABS 2012), equivalent to 253.3 g of CO_2 per km. Average motor vehicle fuel consumption in Australia has remained almost unchanged for more than 50 years. The average fuel economy of passenger vehicles in Australia was 11.4 L/100 km in 1963 (Mees 1999) rising to a high of 12.3 L/100 km in 1991 but remaining at 11.1 L/100 km at the most recent measurement in 2011. The reason for this very modest improvement has been broadly identified by Mees (1999) and Moriarty (1994) as due to a rebound effect such that improved engine efficiency is used to enable larger vehicles to be driven rather than translating to net efficiency gains. In particular the expansion of large sports utility vehicles (SUVs) as a proportion of the vehicle fleet has been a contributor to limited average fuel economy gains.

The fuel economy of new vehicles may be improving over time though this is not entirely certain. Data collected by the National Transport Commission (2014, p. 16) show that the average fuel consumption per kilometre of travel for passenger and light vehicles has declined by approximately 1.5% annually since 2002. A further potential improvement to Australian passenger vehicle fleet fuel economy will likely come from the closure of national automotive manufacturing capacity in 2017. Australian-manufactured vehicles are among the least fuel-efficient within Australian motor vehicle fleets such that their removal from the local market will contribute to overall improvement average fuel economy. Nonetheless there are risks of 'dumping' of inefficient vehicles into the Australian market by international automotive manufacturers. While a voluntary carbon emissions target for new passenger vehicles of 222 g CO_2 per km by 2010 was established by the Australian Federal Chamber of Automotive Industries in 2008, this has not been updated. There are no formal policies to manage the fuel efficiency of Australia's passenger motor vehicle fleets set by the States or the Federal government. Analysis by the Climate Change Authority indicates that Australia is on track for an overall fleet emissions intensity approximately one third higher than the US fleet and double the European intensity. In the absence of national government direction, any transition to more fuel-efficient vehicles will be 'market-led' in that price pressures, such as from higher global petroleum prices, will be the mechanism through which individual consumers receive cost signals as to their behaviour. In turn this implies a distributional question – which urban households will face the greatest pressure to adopt new efficient vehicle technologies and which in turn have the greatest capacity to do so?

13.4 The Distribution of Fuel-Efficient Vehicles

The authors of this chapter have since the late-2000s investigated the distributional aspects of motor vehicle fuel efficiency in cities. These investigations shed light on the potential for a transition in Australian motor vehicle fleets towards greater fuel efficiency and some of the impediments to a rapid transition, including the socio-spatial factors at play. This programme of inquiry has focused on two main questions: first, how is the distribution of the motor vehicle fleet structured in relation to travel patterns in Australian cities and, second, how does this techno-spatial distribution intersect with socio-spatial patterns. Although new vehicle technologies such as electric vehicles are not widely available in Australia yet, the existing petroleum-fuelled motor vehicle fleet can provide indicators as to the uptake of future electric or otherwise fuelled vehicles.

Dodson and Sipe (2007, 2008) demonstrated that Australian suburbia was broadly exposed to differential vulnerability to high global automotive fuel prices. Given contemporary debates about the capacity of fuel-technology shifts to resolve this suburban risk exposure, Dodson Li and Sipe (2009) sought to investigate how Australian motor vehicle fleets are sociotechnically structured both in terms of household socioeconomic status and spatial distribution in terms of vehicle fuel economy. To do this, the authors constructed a dataset of motor vehicle registrations which includes data on the make and model of vehicles registered in Australia including the type, fuel and engine size of the vehicle. This data was matched to the Australian Government's *Green Vehicle Guide* which compiles detailed information on the fuel efficiency of cars available in the Australian market. Because motor vehicle registration data includes the address of the owner, we are able to link the vehicle characteristics to the location of registration, thus enabling spatial analysis of vehicle fleet distribution across cities.

Initial investigations demonstrated that there are distinct spatial patterns in the ownership of motor vehicles among households in Australian cities. For example, in Brisbane the number of vehicles per household tends to be lower in the central city area and higher in the outer and fringe suburban areas (Fig. 13.1). This is not surprising to any observer who is familiar with spatial travel patterns in Australian cities; the relative concentration of land uses and intensity of public transport services, which make non-car travel somewhat more convenient within inner zones, tends to reduce demand for car ownership. In contrast, the post-WWII suburban areas, particularly those developed in the period after the 1970s, tend to be designed around private motor vehicles as the main mode of travel and poorly served by

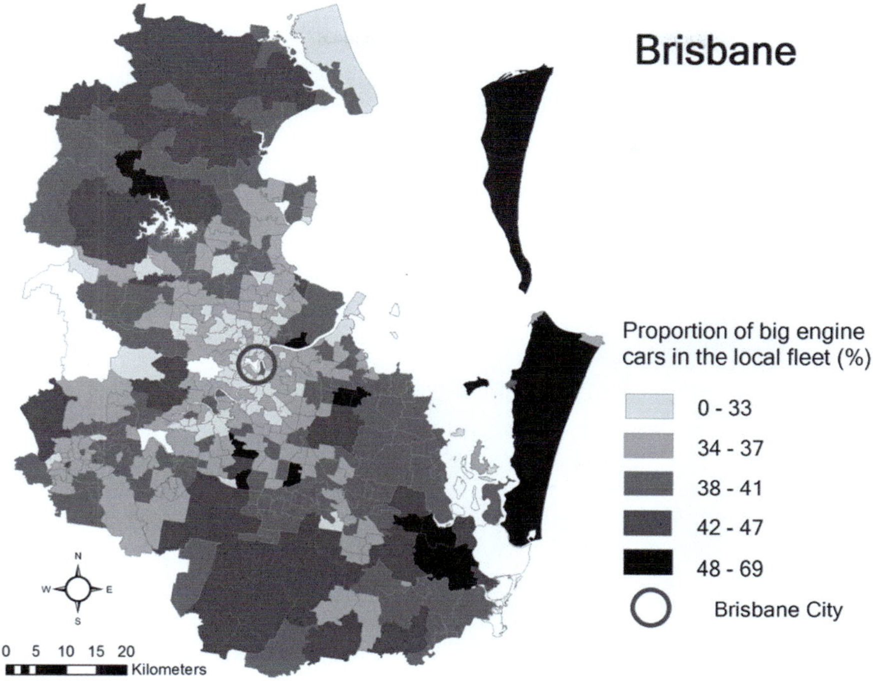

Fig. 13.1 Large engine cars as proportion of the Brisbane passenger vehicle fleet

public transport. It is not surprising that the rate of motor vehicle ownership in these locations tends to be higher than in the inner zones.

While the rate of car ownership in Australian cities tends to show a positive relationship with distance from the city centre, so too does the size of motor vehicles. Australia has a long recognised bias towards six and eight cylinder engine cars, and this is apparent in the distribution of motor vehicles in the Brisbane fleet. As with household car ownership, the proportion of the motor vehicle fleet comprising vehicles of six or more cylinders tends to be associated with location (Fig. 13.1). The inner zones of the city tend to have vehicle fleets with relatively low proportions of six or greater cylinder engines, while engine size tends to increase in the outer and fringe areas.

In contrast to the distribution of vehicles with large engines, the location of newer vehicles, which tend to be of greater fuel economy than older vehicles, is broadly the inverse. The proportion of the vehicle fleet that is no more than 1 year old is around 4% meaning that it would take 25 years for full fleet turnover under prevailing market conditions. The distribution of the new fleet however is highly uneven. In Brisbane the proportion of the motor vehicle fleet which is less than 10 years old tends to be relatively lower in outer and fringe suburban areas than in middle and inner zones (Fig. 13.2). This means that the transition to a more fuel-efficient vehicle fleet is occurring first within inner urban areas and later within outer urban areas.

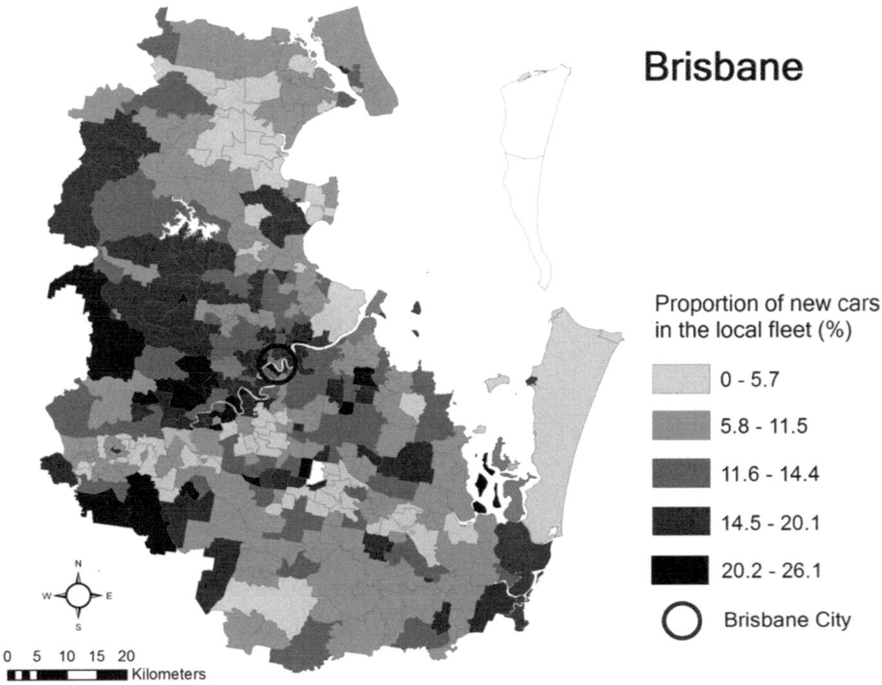

Fig. 13.2 Proportion of passenger vehicle fleet that is less than 1 year old

When considered in total, the patterns of vehicle ownership, engine size and the proportion of new vehicles in Australian cities are clearly spatially differentiated, with more, larger and older vehicles tending to be used in outer suburban areas in contrast to fewer, smaller and newer vehicles being registered to households in middle and inner suburban areas. We are not aware of any immediately identifiable relationship that links vehicle technology to the geographical form of a city in the absence of a social consideration. The technological composition of the Australian urban vehicle fleet is thus implicitly a sociotechnical configuration and moreover one that has socio-spatial dimensions. Indeed it seems appropriate to understand the phenomenon of motor vehicle ownership and deployment in Australian cities as a sociotechnical-spatial configuration. This configuration in turn deserves further elucidation.

13.5 The Socio-spatial Distribution of Vehicles

Further investigations demonstrate that there is a positive association between relatively lower socioeconomic status and the relative inefficiency of motor vehicles. We constructed a spatial model that could differentiate various household characteristics in relation to vehicle fuel efficiency and location, including travel demand. Given Brisbane had been the focus to date, we expanded the analysis to include Sydney. This enabled us to examine how the vehicle fuel efficiency of spatially distributed vehicle fleets varies by household characteristics, including socioeconomic status. To undertake this task required us to combine the motor vehicle registration datasets with the Green Vehicle Guide and in turn to link this to spatial ABS Census data at the suburb level. This is feasible because the registration data and Census data are coded to this relatively disaggregated spatial scale. For each of Brisbane and Sydney, we applied a geographically weighted regression model to the census data including a range of household characteristics at the suburb level, including population size and age, household size, level of employment, housing tenure, commuting mode, level of education, income and housing costs, as well as dwelling density, road density and distance to the CBD.

The results of the geographically weighted regression show that for both Brisbane (Fig. 13.3) and Sydney weekly rent, weekly income and road density tend to show a moderate relationship to vehicle fuel efficiency at the suburb scale, indicating that built environment factors and household socioeconomic position appear to be transferable across both cities. Some variables however apply in one city but not in the other. For example, household size and level of education tend to be more closely associated with higher VFE in Sydney but not in Brisbane, whereas

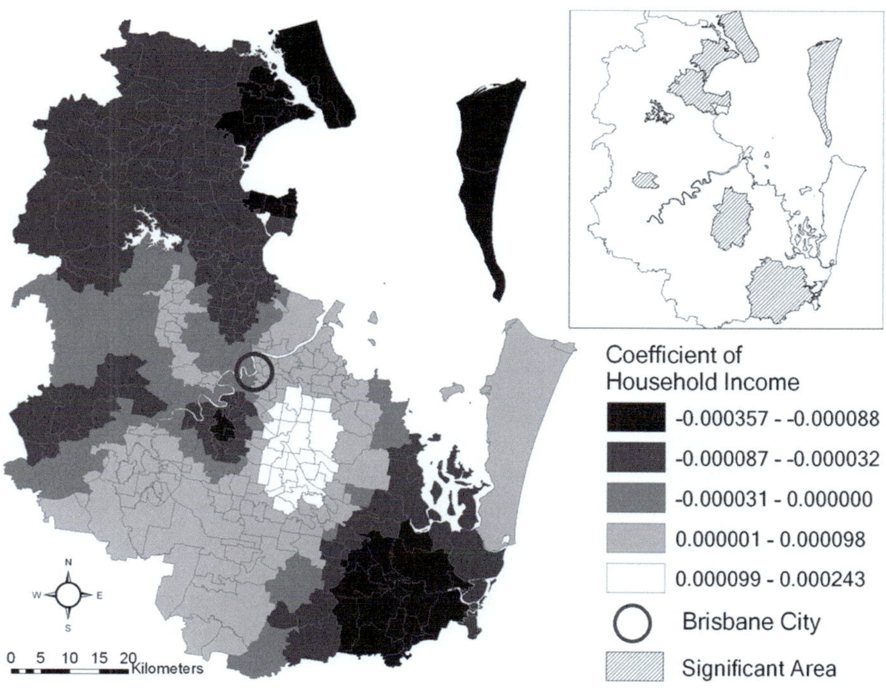

Fig. 13.3 Vehicle fuel efficiency and household income, Brisbane

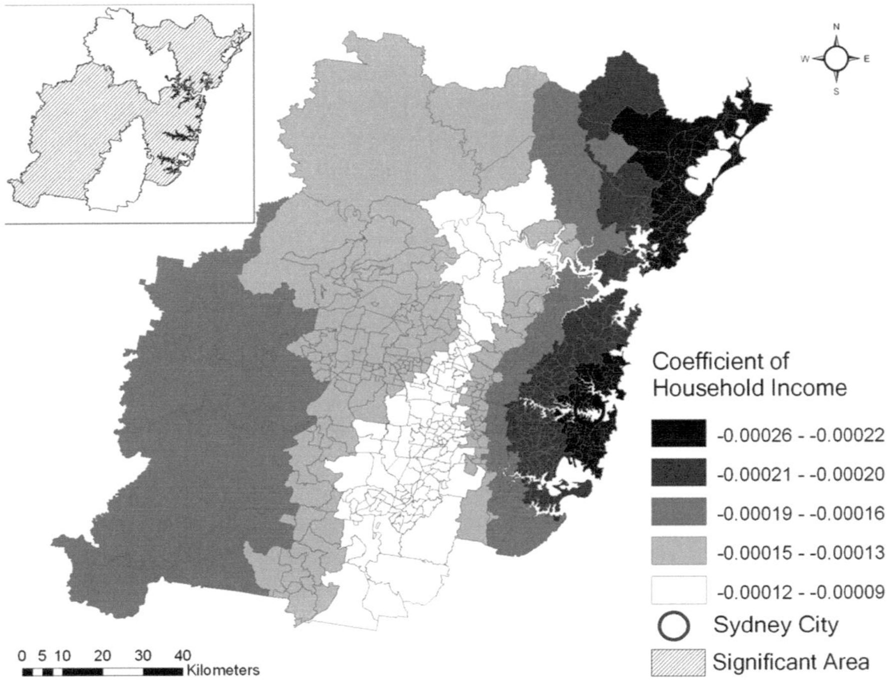

Fig. 13.4 Vehicle fuel efficiency and household income, Sydney

house tenure (owned) and dwelling type (flats) tend to be associated with vehicle fuel efficiency in the latter city. It is perhaps not surprising that larger household size tends to be more associated with larger vehicles. Moreover this analysis shows there is a complex interaction between social, built environment and technological configurations within Australian suburbia (Fig. 13.4).

13.6 Discussion

The expanses of suburbia found across the cities of North America and Australasia are highly dependent on petroleum fuels. The continued supply of petroleum fuels at historically low prices has for nearly a decade been in doubt. Such doubts are only likely to intensify as global reserves of petroleum continue to decline and as global preferences for energy supply shift away from climate harming fossil energy sources. A considerable body of public and policy complimentary assumes that a transition will occur in urban mobility that reduces reliance of individuals and households on conventional fossil-fuelled automobiles. Some European countries, for example, are already considering mandating electric-only vehicle sales within the next decade. In the USA, energy strategy is directly aimed at reducing fuel cost burdens on individual households, while Australia has no mandated fuel efficiency standards.

Suburbia is arguably the urban type that is most problematised by a change in the global petroleum supply environment or a regulatory shift away from fossil fuels, given its dependence on fossil petroleum fuels. The assumption that an easy transition can be made from heavy fossil fuel use for suburban transport to another fuel (and motor) type may be misplaced if those new fuel-efficient vehicles do not filter down quickly through the vehicle market to be available to suburban households. Such households are presently the most dependent at present but purchase new vehicles at relatively lower rates than wealthier households who are also typically located in less car-dependent locations. The consequence of this differential socio-techno-spatial complex is a strong risk of household stress during a transition to non-fossil transport fuels. Policies that rely on market-led adjustments to fleet fuel economy will perpetuate suburban household exposure to future petroleum shocks. Simply hoping for a technology transition is insufficient, given the accompanying social and spatial dimensions of the suburban question. A wider suite of policies, including those that actively redress the spatial infrastructure, service and employment deficits in dispersed suburban areas, must be considered as part of the policy mix to accelerate a transition away from fossil fuels. Policies that do not include such a mix cannot be taken seriously as means to address suburban oil vulnerability.

The problems and stresses facing the suburban automobility complex will no doubt offer fascinating material for scholars of sustainability and technology transitions. In turn however transitions scholars need to develop better theoretical and conceptual models for dealing with suburban questions. This task includes greater integration of the structural underpinnings of suburbanisation in the dynamics of capitalist urban political economy, including the links between land, labour and capital in the context of industrial and post-industrial urbanisation. This urban diagram has been traced by urbanists such as Florida and Walker in the UK and Berry in Australia. For transitions theory to engage with political economy requires a greater attentiveness to the social and political dimensions of transitions rather than applying what has been an often superficial form of analysis that traces in detail the sequencing of technological evolution but only weakly accounts for the social processes at play. The field of transport is highly technological, but it is essential to account for the position of technology within a social and political field and the social solutions that technology permits. In the case of suburbanisation, we must constantly remind ourselves that the provision of transport networks – first rail and then automobiles – was in response to the social pressures of intensive industrialisation and the imperatives for a ‘spatial fix’ for housing stresses. In the case of automobile fuel technology, we need to be attuned to the way the anticipated and emergent transition from fossil fuels to other fuel types reflects the working out of social, environmental and political fissures and tensions, rather than an impersonal process of technological succession.

13.7 Conclusions

Transitions are rarely seamless and linear. So far the question of suburban automobile fuel transitions has barely registered in research, policy and political discussions. Researchers and policy makers are perhaps dazzled by the emergence of non-fossil-fuelled vehicles, such as are being offered by the Tesla Corporation and to a lesser degree by the decisions emerging from some northern European nations to outlaw fossil-fuelled vehicles within the next two decades. It may be that in 20 years motor vehicle fleets will have transitioned from petrol and diesel to electric, hydrogen or some other alternative fuel type. What the experience with high fuel economy vehicles to date shows is that such new technology appears among the higher income social segments first and is then filtered to less affluent households. This recognition necessarily means that any motor vehicle fuel transition, or my systemic transition in motor vehicle technology, will be in the first instance the preserve of affluent urban households. A period of some decades can be expected from the moment of first mass uptake of non-fossil fuel motor vehicles to their saturation as the overwhelming technology form. That multi-decadal period could easily include some serious shocks to the fossil-fuel technological complex, particularly if rapid and comprehensive carbon abatement is adopted by governments globally or if new global geopolitical ruptures impede the supply of oil to car-dependent nations. If such measures or political shifts are experienced, then we can expect considerable economic and social stress to occur within car-dependent suburbia, at a time of emerging social and political disquiet. In turn this recognition demonstrates the need for a transition policy that offers more than market-led fuel adjustment and must take into account wider transformations of urban transport systems. While a fuel-technology transition may shift the reliance of suburbia away from fossil fuels, a more comprehensive mobility transition seems imperative in which a comprehensive reconfiguration of transport systems away from both fossil fuels and automobiles, towards non-fossil-fuelled collective and active modes, such as public transport, walking and cycling, occurs.

References

An F, Earley R, Green-Weiskel L (2011) Global overview on fuel efficiency and motor vehicle emission standards: policy options and perspectives for international cooperation. United Nations Background Paper. UNESDA, Washington, DC

Anable J, Brand C, Tran M, Eyre N (2012) Modelling transport energy demand: a socio-technical approach. Energ Policy 41:125–138

Australian Bureau of Statistics (2012) Survey of motor vehicle use; Cat no. 9208.0; Canberra, Commonwealth of Australia

Banister D (2013) City transport in a post carbon society. In: Givoni M, Banister D (eds) Moving towards low carbon mobility. Edward Elgar Publishing, Cheltenham, pp 255–266

Berry M (1984) The political economy of Australian urbanisation. Prog Plan 22:1–83

Berry M (1999) Unravelling the "Australian housing solution": the post-war years. Hous Theory Soc 16:106–123
Bhat CR, Guo JY (2007) A comprehensive analysis of built environment characteristics on household residential choice and auto ownership levels. Transp Res B Methodol 41(5):506–526
Brand C, Anable J, Tran M (2013) Accelerating the transformation to a low carbon passenger transport system: the role of car purchase taxes, feebates, road taxes and scrappage incentives in the UK. Transp Res A Policy Pract 49:132–148
Broadbent J (1987) The push east: Woolloomooloo Hill, the first suburb. In: Kelly M (ed) Sydney: city of Suburbs. UNSW Press, Sydney
Climate Change Authority (2014) Light vehicle emissions standards for Australia. Canberra, Australian Government
Coenen L, Benneworth P, Truffer B (2012) Toward a spatial perspective on sustainability transitions. Res Policy 41(6):968–979
Cohen MJ (2012) The future of automobile society: a socio-technical transitions perspective. Tech Anal Strat Manag 24(4):377–390
Daly H, Gallachóir BPÓ (2011) Modelling private car energy demand using a technological car stock model. Transp Res Part D: Transp Environ 16(2):93–101
Daly HE, Gallachóir BPÓ (2012) Future energy and emissions policy scenarios in Ireland for private car transport. Energ Policy 51:172–183
Davison G (1979) The rise and fall of marvellous Melbourne. Melbourne University Press, Carlton
Davison G, Yelland S (2004) Car wars: how the car won our hearts and conquered our cities. Allen and Unwin, Melbourne
Delbosc A, Currie G (2013) Causes of youth licensing decline: a synthesis of evidence. Transp Rev 33(3):271–290
Dennis K, Urry J (2013) After the car. Wiley, New York
Dodson J (2012) Transforming Australia's housing solution: how we can better plan suburbia to meet our future challenges. In: Tomlinson R (ed) Australia's unintended cities: the impact of housing on urban development. CSIRO Publishing, Melbourne
Dodson J (2014) Suburbia under an energy transition: a socio-technical perspective. Urban Stud 51(7):1487–1505
Dodson J, Sipe N (2007) Oil vulnerability in the Australian city: assessing socio-economic risks from higher urban fuel prices. Urban Stud 44(March):37–62
Dodson J, Sipe NG (2008) Shocking the suburbs: oil vulnerability in the Australian city. UNSW Press, Sydney
Dodson J, Sipe N (2009) A suburban crisis?: housing, credit, energy and transport. J Aust Polit Econ 64:199
Florida R, Jonas A (1991) U.S. urban policy: the postwar state and capitalist regulation. Antipode 23(4):349–384
Frost L (2000) Connections. In: Troy P (ed) A history of European Housing in Australia. Cambridge University Press, Cambridge
Gallagher KS, Muehlegger E (2011) Giving green to get green? Incentives and consumer adoption of hybrid vehicle technology. J Environ Econ Manag 61(1):1–15
Geels DIFW (2005). The dynamics of transitions in socio-technical systems: a multi-level analysis of the transition pathway from horse-drawn carriages to automobiles (1860'Äì1930). Tech Anal Strat Manag 17(4): 445–476
Geels FW, Schot J (2007) Typology of sociotechnical transition pathways. Res Policy 36(3):399–417
Gilbert R, Perl A (2008) Transport revolutions: moving people and freight without oil. Earthscan, James & James, Hoboken

Glerum A, Frejinger E, Karlstrom A, Beser Hugosson M, Bierlaire M (2013) Modeling car ownership and usage: a dynamic discrete-continuous choice modeling approach. International choice modelling conference
Gonzalez GA (2009) Urban sprawl, global warming, and the empire of capital. SUNY Press, Albany
Goodwin P (2013) Peak travel, peak car and the future of mobility: evidence, unresolved issues, policy implications. ITF Round Tables Long-run Trends Car Use 152:53
Graham-Rowe E, Skippon S, Gardner B, Abraham C (2011) Can we reduce car use and, if so, how? A review of available evidence. Transp Res A Policy Pract 45(5):401–418
Grimal R, Collet R, Madre J-L (2013) Is the stagnation of individual car travel a general phenomenon in France? A time-series analysis by zone of residence and standard of living. Transp Rev 33(3):291–309
Guo JY, Bhat CR (2007) Operationalizing the concept of neighborhood: application to residential location choice analysis. J Transp Geogr 15(1):31–45
Harvey D (1985) The urbanization of capital: studies in the history and theory of capitalist urbanization. John Hopkins University Press, Baltimore
Headicar P (2013) The changing spatial distribution of the population in England: its nature and significance for 'peak car'. Transp Rev 33(3):310–324
Hirschhorn JS (2005) Sprawl kills: how Blandburbs steal your time, health and money. Sterling & Ross Publishers, New York
Hulsmannm M (2012) General sales forecast models for automobile markets and their analysis. MLDM 5(2):65–86
Jackson K (1987) Crabgrass frontier: the suburbanization of the United States. Oxford University Press, Boston
Kagawa S, Nansai K, Kondo Y, Hubacek K, Suh S, Minx J, Kudoh Y, Tasaki T, Nakamura S (2011) Role of motor vehicle lifetime extension in climate change policy. Environ Sci Technol 45(4):1184–1191
Kloess M, Müller A (2011) Simulating the impact of policy, energy prices and technological progress on the passenger car fleet in Austria—a model based analysis 2010–2050. Energ Policy 39(9):5045–5062
Kuhnimhof T, Zumkeller D, Chlond B (2013) Who made peak car, and how? A breakdown of trends over four decades in four countries. Transp Rev 33(3):325–342
Li T, Sipe N, Dodson J (2013) Investigating private motorised travel and vehicle fleet efficiency: using new data and methods to reveal socio-spatial patterns in Brisbane, Australia. Geogr Res 51(3):269–278
Liddle B (2012) The systemic, long-run relation among gasoline demand, gasoline price, income, and vehicle ownership in OECD countries: evidence from panel cointegration and causality modeling. Transp Res Part D: Transp Environ 17(4):327–331
Lipschutz RD (2012) Getting out of the CAR: decarbonisation, climate change and sustainable society. Int J Sustain Soc 4(4):336–356
Liu J, Santos G (2013) Decarbonising the road transport sector: breakeven point and consequent potential consumers' behaviour for the US case. Int J Sustain Transp 9(3):–159, 175
Mees P (1999) A very public solution: transport in the dispersed city. Melbourne University Press, Melbourne
Mees P (2010) Transport for suburbia: beyond the automobile age. Earthscan, London
Metz D (2010) Saturation of demand for daily travel. Transp Rev 30(5):659–674
Metz D (2013) Peak car and beyond: the fourth era of travel. Transp Rev 33(3):255–270
Millard-Ball A, Schipper L (2011) Are we reaching peak travel? Trends in passenger transport in eight industrialized countries. Transp Rev 31(3):357–378
Moriarty P (1994) Can alternative car fuels reduce greenhouse gas emissions? Int J Veh Des 15(1-2):1–7
National Transport Commission (2014) Carbon dioxide emissions from new Australian vehicles 2013. Melbourne, National Transport Commission

Newman P, Kenworthy J (1999) Sustainability and cities: overcoming automobile dependence. Island Press, Washington, DC

Puentes R, Tomer A (2008) The road less traveled: an analysis of vehicle miles traveled trends in the US. Washington, DC, Brookings Institution

Rigal A, Rudler J (2014) Post-car world: why laugh at change? 14th Swiss transport research conference. Monte Verita, Ascona

Schipper L, Deakin E, Sperling D (2010) Sustainable transportation: The future of the automobile in an environmentally constrained world. University of California Transportation Center

Schiraldi P (2011) Automobile replacement: a dynamic structural approach. RAND J Econ 42(2):266–291

Sheinbaum-Pardo C, Chávez-Baeza C (2011) Fuel economy of new passenger cars in Mexico: trends from 1988 to 2008 and prospects. Energ Policy 39(12):8153–8162

The White House (2011) Blueprint for a secure energy future. The White House, Washington, DC

Tyfield D, Urry J (2012) Greening China's 'Cars': could the last be first? Lancaster, Centre for Mobilities Research, Lancaster

Urry J (2012) Do Mobile Lives Have a Future? Tijdschr Econ Soc Geogr 103(5):566–576

van Dender K, Clever M (2013) Recent trends in car usage in advanced economies–slower growth ahead? International Transport Forum, OECD, Paris

Walker R (1981) Theory of suburbanization: capitalism and the construction of the urban space in the United States. In: Dear M, Scott A (eds) Urbanisation and urban planning in capitalist society. Methuen, London/New York

Winter I, Bryson L (1998) Economic restructuring and state intervention in Holdenist suburbia: understanding urban poverty in Australia. Int J of Urban Reg Res 22(1):60–75

Part V
Conclusions

This final chapter summarises the key contributions from the book – highlighting that it brought together transitionists and urbanists around a shared interest and in the Australian context. It discusses their implications for future research and practice of urban sustainability transitions in Australia and internationally.

Chapter 14
Urban Sustainability Transitions: An Emerging Hybrid Research Agenda

Ralph Horne, Trivess Moore, Fjalar de Haan, and Brendan James Gleeson

Abstract This chapter takes stock of the fourfold ambition of this book, namely:

- To introduce transitions scholars and practitioners to urban studies
- To introduce urban scholars and practitioners to transitions studies
- To collect and present case studies based in Australian cities that intersect urban and transitions themes and present these in a global setting of climate emergency and urbanization
- To introduce a wider, global audience to urban transitions ideas, scholarship and practice as it is emerging in various ways across Australia.

It also introduces the Australia-based Sustainability Transitions Researchers Alliance (ASTRA) as the network that provided the basis for the book. In reflecting on the above ambition we pose the rhetorical question: "How can urban and transitions perspectives assist understandings of change towards sustainability using examples relevant to Australian cities?" In answer, we describe how urban studies can be enriched by the exuberance offered by a newly emerging field of transition studies. In turn, we show how transition studies can benefit from spatial, political and urban perspectives of dynamic city processes. The perspectives presented throughout the book are placed in the context of contemporary urbanization, in a still unfolding post-Global Financial Crisis era when increasing questions are being asked of globalization, neo-liberalism, and the unequal and unsustainable outcomes of urban governance processes.

R. Horne (✉)
College of Design and Social Context, RMIT University, Melbourne, VIC, Australia
e-mail: ralph.horne@rmit.edu.au

T. Moore
School of Property, Construction and Project Management, RMIT University, Melbourne, VIC, Australia
e-mail: trivess.moore@rmit.edu.au

F. de Haan • B.J. Gleeson
Melbourne Sustainable Society Institute and Melbourne School of Design, The University of Melbourne, Melbourne, VIC, Australia
e-mail: fjalar.dehaan@unimelb.edu.au; brendan.gleeson@unimelb.edu.au

T. Moore et al. (eds.), *Urban Sustainability Transitions*, Theory and Practice of Urban Sustainability Transitions, DOI 10.1007/978-981-10-4792-3_14

Keywords Urban studies • Transition • Australia • ASTRA • Sustainability transitions • Transition Management

Climate change and the rush to cities are set to be defining challenges of our time, with major implications for global security, non-human species sustainability and the health and wellbeing of the human race. What is patently needed is an urgent and significant transition away from fossil-based, energy-intensive, consumption-driven cities. However, how, who, what, when, where and with what consequence are current and potential future efforts in this regard positioned? As indicated in the introductory chapter, our ambition in this book has been fourfold:

- To introduce transitions scholars and practitioners to urban studies
- To introduce urban scholars and practitioners to transitions studies
- To collect and present case studies based in Australian cities that intersect urban and transitions themes and present these in a global setting of climate emergency and urbanisation
- To introduce a wider, global audience to urban transitions ideas, scholarship and practice as it is emerging in various ways across Australia

In taking transitions scholars on a tour of the city, we aim to not only describe how cities work but also to seek ways to highlight key challenges in cities and in our understanding of cities. At risk of overgeneralising, urban scholars tend to emphasise spatial, political-economic and power dynamics in the obduracy and dynamics of cities. They tend to view cities as people, rather than technologies, where the struggle for resources, including space, plays out constantly across the city – that is, across the extant materiality that represents the results of previous struggles.

In taking urban scholars on an outing through (in particular) a variety of interpretations of the emerging field of transitions studies, we aim to illustrate both the youth and exuberance of a topical genre and also the possibilities for enriching our understanding of cities that this presents. While more purposive interpretations of transitions studies might appear rather limiting for studies of city governance and urban change, they may help shed new light upon components of urban change. Meanwhile, detailed empirical studies of socio-material change have immediately recognisable value to bring to those of us concerned with ideas of low carbon cities and how they may be configured in an era of late neo-liberal, globalised networks facing climate shock.

With these narratives running along in the background, each chapter is deliberately quite independently organised in order to provide a multiplicity of perspectives, topics, scales, underlying theories and emphases. What binds them together is the triple-strand thread of sustainability transition, urban settings and Australia. If there is a guiding question, it would be something like: *How can urban and transitions perspectives assist understandings of* change towards sustainability, *using examples relevant to Australian cities?* However, quite intentionally, there is no single question, and contributors have been encouraged to take their own urban journey and see where it leads to.

Part I provides an introductory and overarching set of chapters, offering critical-comparative, conceptual and political perspectives. Some prefer an urban theoretical sensibility, others a transitions studies one. Chapter 1 by Horne sketches out key literature and concepts of both as they relate to low carbon futures. It challenges both transitions and urban studies to highlight the 'who and how' of attempts at purposive steering of urban change. It also poses questions about how cities, literally built from fossil exploitation and highly dependent in their current form upon the associated affordances of consumption, might be capable of contributing to a shift away from carbon and, if so, what other structures would need to synchronise or align. It also positions Australia as a site for relevant studies and presents key ideas of spatial, power and socio-material 'ways in' to contemplating low carbon housing and urban change. By way of practical examples, it explores the limits and utility of ideas of socio-technical niches.

Chapter 2 by de Haan is situated in the diverse transitions studies genre and analyses the conceptual reasons for the field's issues in dealing with matters of place and scale. It provides a way to probe how much place can matter for a transition: is it *essentially* place based or merely *accidentally* so? If place matters, it can be conceptualised as a *locus* or as a *nexus* – providing a frame for a geographically sensitive treatment. In stark contrast, Gleeson in Chap. 3 offers a quintessentially urban perspective. It contemplates the inevitable major shifts that significant realignment of carbon-related capitalism will bring and suggests a two-stage process is likely if we are to survive and provide humanity with a more sustainable future. These are shaped around political projects and power relations of urban transition.

Part II presented a collection of four discrete urban sustainability transitions accounts, all sharing the same meta-theme – namely, the prospects for governing purposive transitions – yet each drawing from distinct traditions (two more urban centred, two more transitions centred) and in so doing emphasising different elements and dynamics. Chapter 4 by Morrissey et al. offers a critical spatial planning (urban) perspective and contrasts two cases in two countries – Ireland and Australia – as a means to reveal the primacy of space, place and political context in transitions studies. Chapter 5 by Bush et al. adopts a practitioner stance on sustainability transitions, and the resultant narrative offers an intimate insider account of what we might call niche organisations navigating their way through uncertain urban transitions territory, aided and abetted by meaningful links with scholarship and research. This offers sense-making in real time in ways that are rarely afforded practitioner communities.

Moloney et al. in Chap. 6 extend the concept of 'boundary organisation' to also include other boundaries apart from the science-policy one, to delineate, situate and sort sustainability transitions at the local and regional level. Boundaries in this case are variously spatial, social, cultural, jurisdictional, scalar, organisational and institutional, and they in turn generate new challenges for both transitions studies and contemporary governance. In Chap. 7 Moore focuses on Strategic Niche Management, drawn from transitions studies, and seeks through examples and explanations as to why some niche projects appear to be more successful than others, both in scale and impact on the broader regime.

Part III presents three popular topics in urban change and offers transitions study-based interpretations. Chapter 8 by Brown et al. is about the process of coalition building, in this case, towards water-sensitive cities, while Chap. 9 (Newton) adopts a Transition Management schema as a way to illustrate what is missing from successful and sustainable conversion of Melbourne's greyfields. In Chap. 10 Alexander and Rutherford offer a transitions study frame to the idea of the anti-capitalist urban movement in an optimistic rendition of the prospects for the Transition Towns campaign.

The penultimate Part IV offers three overtly 'spatial' contributions, in that, while spatial dimensions are critical, they each also tackle complex urban governance challenges of transitions. Chapter 11 by Dalton focuses on suburban development and, within this phenomenon, how the underlying 'lock-in mechanisms' producing and reproducing the suburbs have at times been destabilised and reconfigured. This historical view is deliberately presented as a way of contemplating the prospects for future change. Doyon in Chap. 12 examines changing work spaces and the associated enabling/disabling social, technical and governance structures. Chapter 13 by Dodson et al. turns to urban mobility and contemplates the possibility of further segregative and inequality effects from market-led shifts to highly efficient motor vehicles.

In all, this edited volume has explored urban sustainability transitions across a range of different transitions perspectives. It builds upon the emerging stream of urban transitions research, including the preceding book in this series by Loorbach et al. (2016). This volume explores cases which are emergent rather than purposive to provide another layer of understanding about urban transitions. Furthermore this is not a book of transitionists looking at the urban but brings together a mix of transitionists and urbanists. It aligns with and extends debates about the spatial turn in transitions studies with considerations of scale and place, a contrast to most of the transitions literature to date. In this way, chapters by authors such as de Haan, Morrissey et al., Doyon, etc. are helping to further this important but undervalued element within urban transitions studies.

This book is also a showcase of the Australian research on transitions. As mentioned in the introduction, this volume is the early fruit of the Australia-based Sustainability Transitions Researchers Alliance or ASTRA. This network of researchers first came together in early 2015, but transitions research in Australia goes back quite a bit further. A characteristic of Australia-based transitions research to date – though this may very well change – is that there are no transitions institutes and few dedicated transitions research groups. In other words, these researchers are often separated from their transitions colleagues, which in Australia can mean they are time zones apart. ASTRA was founded to overcome this and be a platform for research exchange, collaboration and communication. ASTRA hosts seminars and reading group sessions and produced this book.

This book may however wrongly raise the impression that Australian transitions researchers are only interested in urban issues. This is surely not the case. Energy transitions and low carbon futures – as anywhere – are an important concern, and these are not particularly confined to the urban. Nevertheless, there is a lot of attention to the urban in Australian transitions research, and perhaps this is a reflection of Australia being a very urban nation (see Introduction and Chap. 1). Perhaps this context makes Australian transitions research particularly relevant.

It seems, just as urbanisation is speeding up, so are political shifts and the number of well-intentioned urban responses. President Trump, Brexit and a wave of reaction to the neo-liberal globalisation project are adding considerable momentum to the questions raised by the likes of Thomas Piketty in the wake of the global financial crisis. At the same time, UN Habitat have launched the 20-year plan for sustainable and equitable cities under the New Urban Agenda. This follows the launch of the UN Sustainable Development Goals with a new emphasis on sustainable cities.

It is clear from this book series that there is a range of complex challenges we face if a transition to a more sustainable urban future is to be achieved. Changing the deep structural issues embedded within existing regimes is not easy or quick to achieve, as explored throughout the broader sustainability transitions literature. Such changes may take generations to achieve and time which we may not have, especially if we are to avoid catastrophic climate change. As regards the learnings the different volumes in this book series present for practitioners and policymakers, we can already see a spectrum of contributions emerge. The inaugural volume by Loorbach et al. (2016) derives lessons for what may become a more specific 'Urban Transition Management', and it can do so in considerable detail and depth because it has a consistent focus on cases where Transition Management is implemented. The value of this volume, by contrast, lies in providing a kaleidoscopic overview of the intersection of transitions efforts and research with urban governance and policy. For practitioners and policymakers, this provides a broad introduction to how the urban issues they work on are transitions issues and the various implications of this. For transitions scholars, it provides insights into the way urban scholars understand urban change. Neither perspective is preferenced; indeed, the value of this book is in critical engagement with both in the spirit of recognising the strengths and weaknesses each can bring to the urgent task of urban transitioning.

As the chapters throughout this book have demonstrated, there is an emerging engagement in Australia by urban researchers and practitioners with how to transition to a more sustainable and equitable future. This engagement is raising questions about the fate of ecological modernisation, about the role of capitalism in both carbon fuelled cities and post-carbon fuelled cities and about the crossovers and mutual ground shared between environmental, inequality and governance questions facing cities. This is, of course, a burgeoning and urgent agenda, and this book is inadequate to the gargantuan task. Our ambition is to highlight certain aspects. In a future book, we hope to add more ideas to progress this critical debate.

Reference

Loorbach D, Wittmayer JM, Shiroyama H, Fujino J, Mizuguchi S (eds) (2016) Governance of urban sustainability transitions. European and Asian experiences. Springer Japan, Tokyo. https://doi.org/10.1007/978-4-431-55426-4

Index

T. Moore et al. (eds.), *Urban Sustainability Transitions*, Theory and Practice of Urban Sustainability Transitions, DOI 10.1007/978-981-10-4792-3

MIX
Papier aus verantwortungsvollen Quellen
Paper from responsible sources
FSC® C105338

FSC
www.fsc.org

If you have any concerns about our products,
you can contact us on
ProductSafety@springernature.com

In case Publisher is established outside the EU,
the EU authorized representative is:
Springer Nature Customer Service Center GmbH
Europaplatz 3, 69115 Heidelberg, Germany

Printed by Libri Plureos GmbH
in Hamburg, Germany